The Distraction Trap

The Distraction Trap

How to focus in a digital world

FRANCES BOOTH

PEARSON

Harlow, England • London • New York • Boston • San Francisco • Toronto • Sydney
Auckland • Singapore • Hong Kong • Tokyo • Seoul • Taipei • New Delhi
Cape Town • São Paulo • Mexico City • Madrid • Amsterdam • Munich • Paris • Milan

PEARSON EDUCATION LIMITED

Edinburgh Gate
Harlow CM20 2JE
United Kingdom
Tel: +44 (0)1279 623623
Fax: +44 (0)1279 431059
Web: www.pearson.com/uk

First edition published 2013 (print and electronic)

ISBN: 978-0-273-78585-9 (print)
978-0-273-78860-7 (PDF)
978-0-273-78859-1 (ePub)

British Library Cataloguing-in-Publication Data
A catalogue record for the print edition is available from the British Library

Library of Congress Cataloging-in-Publication Data
Booth, Frances.
The distraction trap : how to focus in a digital world / Frances Booth. -- First edition.
pages cm
Includes bibliographical references.
ISBN 978-0-273-78585-9 (limp)
1. Technological innovations--Psychological aspects. 2. Technological innovations--Social aspects. 3. Information technology--Psychological aspects. 4. Information technology--Social aspects. 5. Distraction (Psychology) I. Title.
HM846.B66 2013
303.48'33--dc23
2013000758

10 9 8 7 6 5 4 3 2 1
17 16 15 14 13

Cover design by Two Associates
Illustrations by Stu McLellan
Print edition typeset in 11/14pt Sabon by 3
Print edition printed and bound in Great Britain by Clays Ltd, Bungay, Suffolk

Note that any page cross references refer to the print edition

To Grandma and in memory of Grandad.
Thank you for believing in me.

Contents

About the author

FRANCES BOOTH STUDIED SOCIAL AND Political Sciences at Cambridge University and Journalism (MA) at Sheffield University, before working for eight years as a journalist for *The Daily Telegraph* and *The Guardian*. When she noticed digital distraction all around her getting worse, she decided to write this book.

Frances provides digital distraction consultancy for individuals and businesses, and speaks at events. She also runs *Here are some words*, providing training for other writers and passing on an enthusiasm for words.

Frances grew up in Newcastle upon Tyne in a time before people had even heard of email. She now works among tech-savvy entrepreneurs in an innovation space in Tech City, London.

www.thedistractiontrap.com
www.herearesomewords.com

Acknowledgements

TO EVERYONE WHO HAS EVER said to me 'How's the book?' thank you. Your encouragement and enthusiasm kept me going so many times.

To my Mum and Dad, Judith and Mike Booth, thank you for your never-ending support, love and encouragement, and for sharing this whole journey with me.

Thank you, Mum, for showing me great writing is just an everyday thing, and thank you, Dad, for a lifetime's worth of excellent advice.

To my agent, Jacq Burns, thank you for believing in this idea, for pushing me to make it better, and for being great fun to work with.

To Sam Jackson at Pearson, thank you for your spot-on suggestions that made this book better. Thanks also to everyone else at Pearson who worked on this book.

To Nathalie Nahai, thanks for your sound advice as I set off on the path you were just coming to the end of, and thanks to Sarah Lloyd-Hughes for a generous introduction that helped this book get started.

To my incredible brother, David Booth, whose no-nonsense advice I have always been able to rely on, thank you for sharing my huge excitement about this book. Thanks also to Lulu Chen for your pure enthusiasm.

To my cousin, Sally Hamlyn, thank you for cheering me on all the way, and thank you, too (and to Dan, Imogen and Miles Maclaren) for my 'country retreat' I always knew I could escape to. Thank you also to my aunty, Pat Wilson, for having such belief in me as a writer.

To Andrea Case, thank you for sharing my excitement about this book before anyone else, and for years of incredible friendship.

To Sally Priestley, thank you for being there to discuss (at length!) all things in life, and particularly for your ideas on relaxation.

Thank you to Homa Khaleeli, Guy Perera, Yolima Olaya, Rebecca Davies, Jessica Mills, Luke Fairless, Ffiona Kyte, Susie Armstrong, Catherine Humble and John Burke for having such enthusiasm for this book and for my writing.

To Araceli Camargo, thank you for pushing me to aim higher and higher. To Sarah Colthorpe, thank you for your unending enthusiasm for all my creative endeavours, and for being such a bright, creative spark.

A huge thank you to everyone on the London Cuban salsa scene. Dancing kept me going as I wrote this book. Many projects brought me to this point as a writer. Thanks particularly to Catherine Warrington for being a wise guide, and to Helene Dancer for sharing a fun-filled writing adventure.

To everyone who told me a story, sent me a link, or discussed digital distraction with me while I wrote this book, thank you.

I thoroughly enjoyed those conversations. They were invaluable and sparked ideas at every stage. Thank you particularly to my case studies for sharing their stories with everyone.

Thanks to you, finally, for reading. I hope you continue the debate.

PUBLISHER'S ACKNOWLEDGEMENT

We are grateful to the following for permission to reproduce copyright material:

Table

Table on pages 62–4 from *Caught in the Net*, John Wiley (Young, Kimberly S. 1998) pp. 31–3. Copyright © 1998 Kimberly S. Young. Reproduced with permission of John Wiley & Sons Inc.

In some instances we have been unable to trace the owners of copyright material, and we would appreciate any information that would enable us to do so.

How to use this book

DO YOU FEEL AS IF YOU never have any time to yourself these days?

As though you are always 'on call' to emails, text messages and status updates?

This book will show you step by step how to tackle digital distraction.

If it's someone close to you who is digitally distracted, it will help you coax them away from a screen and make them pay attention to you.

Attention is the pot of gold we're searching for.

Perhaps you've noticed that email, smartphones, social media and the Internet have been demanding a bit much of that recently.

It's time to win some back – to give to each other and to the projects that mean the most to us.

What happens when you are focused rather than distracted?

You're more productive, less stressed, and that incessant background buzz is… somehow gone.

Tempted to try it?

There are nine steps to getting focused.

The first step is to assess your usage (most of us have become a lot more digitally distracted than we think).

Then, you'll learn how to identify what a feeling of focus means for you.

You'll take tips from the pros (many of those working in technology are early adopters of good habits).

You'll also get a handy list of 15 tips to use when you are distracting yourself.

After that, you'll go on to look at what – and who – is distracting you.

In the final chapters, you'll learn how to aim high, and how to keep balanced and focused, whatever challenges appear.

The practical section of the book (the nine steps) is Part 2. In Part 1, you'll learn about the background to the problem.

The book is based on research, publications and studies in the fields of distraction, attention and digital life. It is also based on observations and many, many conversations I have had about digital distraction.

Everyone I spoke to about this book had a story to tell me. These conversations were hugely valuable and I thank all of you who took the time to share with me a story about how digital distraction was changing life for you.

Eight of these stories are told in depth as case studies – you'll find them as you go through Part 2. They include the digitally distracted, as well as those who have come up with innovative ways to combat distraction. Some names have been changed to allow them to share their stories.

How should you read this book?

The most important thing is to allow yourself space to take the information in by switching off all digital distractions (your phone, email, the Internet and social media) each time you read a chapter.

Each chapter should not take too long to read.

When you've finished each chapter, switch back on.

This might seem strange or uncomfortable at first. Try it just once as an experiment and go from there.

Feel free to make notes all over the book – there'll be questions along the way, and lots to think about. Perhaps underline things that mean something to you, so you can spot these sections easily.

It's also a good idea to get a notebook to make notes in. This is your 'digital distraction notebook'.

Here, you can keep track of your thoughts, answer questions, and check on your progress at the end of each chapter. There'll be a progress check at the end of each chapter in Part 2.

Each chapter builds on the one before, so the book is best read from front to back. If you're really struggling at the start, just skip straight to Part 2 and get practical.

There is a full list of references at the back of the book if you want to do further research. There is also a recap at the end of each chapter, so you can remind yourself what you've just read.

You can expect to start noticing results by around the middle of Part 2. By the end of the book, the aim is to be in control of how you use digital platforms, rather than the other way round.

One final thing – don't keep this book to yourself. Ask people around you what they think. Perhaps they're experiencing very similar issues to you.

Now, if you're ready, go and gather every digital device you own – including laptops, smartphones, mobiles, tablets, BlackBerries, iPads, iPhones and computers, and put them all in a pile somewhere you can see them.

Done that? Great.

Just so you know what you're facing.

Now – if you can bear it – turn them off for five minutes… and read on.

PART 1

Distraction

---

noun something that distracts one's attention

CHAPTER 1

The way we live

If you're in the room, be in the room. NIGEL RISNER

THE SMELL OF COOKING IS coming from the kitchen, as an easy conversation flows back and forth. It is a summer evening, a rare chance for some quality time. Beep. Beep. There goes the smartphone. For some reason, you immediately go to it.

'Hmm... what was that?' you say, not listening, as you check your email (which isn't important), and notice another email (that could wait), and allow your attention to wander off. You start checking through all your emails. Then, to answer something, you head to Google. It'll only take a minute. And while you're there...

What about your conversation? Where might it have gone? What about your memories of that evening? They, too, have been sabotaged.

What about the other person in the room?

Digital distraction means our behaviour has suddenly changed. Without noticing it, we've become caught in a trap.

We're damaging our relationships, increasing our stress levels, and literally rewiring our brains. We are convinced we can do 10 things at once – it all seems so high-speed. But the reality is, we're failing to get anything done. We're constantly overwhelmed, we never have time, and we're forgetting how to do simple tasks, like read.

In short, we are losing the ability to pay meaningful attention to anyone or anything.

How did it get to this?

LIVING HISTORY

Rewind about 1,000 years or so to 1997. A search engine called BackRub – still in its early stages – is looking for a new name. 'Google?' someone suggests. And we begin living history…

It's also about now that we start using email as a widespread way to communicate with each other (before that, academics, government and the military used it). One of the first web-based email services, hotmail, launches in 1996. In 1998, the word 'spam' is added to the *Oxford English Dictionary*.[1]

Next, we start to go mobile. In 1999, BlackBerry launches. By 2004, we're checking email on the move in our millions.[2]

The same year (drum roll, please…) along comes Facebook. We embrace social media with unending enthusiasm. In 2006, Twitter decides to join in. Then, in 2007, the first iPhone arrives.

Did all that happen in just a decade?

As Ian Price says: 'It might feel as if email and BlackBerry have crept into our lives gradually, but in the context of a 5,000-year history of written communication, their arrival has been sudden and disruptive.'[3]

If you've felt like you've been treading water just to keep up, now you know why.

We haven't even stopped to catch our breath. We've just embraced it and embraced it some more...

And look at what we can do! We're experts; way beyond finding the 'on' button now. We can customise, build our own programmes, and create virtual worlds on the move.

Yes, we know how to use technology – but do we actually know the best way to fit it in to our lives?

We don't really need statistics to tell us what's happening.

We can feel it. We can see it. We sit on a train and notice everyone is staring down at a handheld screen. We sit in our homes and notice no one is talking to each other – they're all texting, instant messaging, or Skyping other people. We sit in our offices and notice everyone is battling against a never-ending information stream.

To support these observations, solid research has begun to emerge, of addiction, of lack of productivity, and of people feeling overwhelmed.

This year, for the first time, Internet Addiction will be included in the manual used worldwide to classify mental disorders – the *Diagnostic and Statistical Manual of Mental Disorders* (DSM). It will be marked 'for further study'.

It's time we took a closer look at the state we're in.

THE STATE OUR BRAINS ARE IN

There's no doubt about it. We're wired.

We have wires coming out of our ears, wires coming out of our pockets, wires trailing out of our handbags. In the office, on the street, on the train, you see them – wired people.

It's not unusual to be on your smartphone from the moment you wake up in the morning (in fact it wakes you up) to the moment you go to sleep at night (when your thoughts turn to the emails you've just read in bed).

We used to wake to the sound of birds. Now it's message alerts.

And, as a result of this constant connectivity, something shocking has happened – we have literally rewired our brains.

The fact that the brain changes is nothing new (for example, when we learn something like how to play a musical instrument). Scientists call the malleability of the brain neuroplasticity.

When we learn a new skill like playing an instrument, connections are formed between neurons, making new pathways within the brain.[4] If we then stopped playing the instrument, for example, the links wouldn't be in use, and would weaken. New strong connections would be formed elsewhere – based on whatever we were doing.

While we are learning and new pathways are forming, things feel difficult. Once strong new pathways are established and behaviour is almost automatic, it seems easy.

Think of how easily – without even noticing it – you log in to your email or social media account. It's automatic. An extremely strong brain pathway has been established.

In his excellent, in-depth book on this, *The Shallows*, Nicholas Carr explains why the pull of the Internet is so strong:

> If, knowing what we know today about the brain's plasticity, you were to set out to invent a medium that would rewire our mental circuits as quickly and thoroughly as possible, you would probably end up designing something that works a lot like the Internet.

He says that the Internet delivers 'precisely the kind of sensory and cognitive stimuli – repetitive, intensive, interactive, addictive – that have been shown to result in strong and rapid alterations in brain circuits and functions.'

The new pathways that we have created, without even meaning to, say 'check email', 'Tweet', 'Google', 'check Facebook'. These pathways are now very, very strong, and it's hard not to travel down them. In most cases, we reinforce these pathways every day. In some cases, we reinforce them every 10 minutes.

What is alarming, is that we have changed our brains so quickly, and so much, without even really realising it.

Scientist Gary Small and his team studied the changes that happen to the brain when we use the Internet. They found that, after just five hours of Internet use, a person's brain began to change.[5]

Another study in China, of a group of 19-year-olds who spent an average of 10 hours a day in front of a screen, showed that excessive use of the Internet is associated with shrinkage of certain parts of the brain. The changes were in the parts of the brain that control attention and emotional processing.[6]

While we are building new 'digital pathways', old pathways are being neglected.

As Carr explains:[7]

> Just as neurons that fire together wire together, neurons that don't fire together don't wire together. As the time we spend scanning Web pages crowds out the time we spend reading books, as the time we spend exchanging bite-sized text messages crowds out the time we spend composing sentences and paragraphs... the circuits that support those old intellectual functions and pursuits weaken

and begin to break apart. The brain recycles the disused neurons and synapses for other, more pressing work. We gain new skills and perspectives but lose old ones.

SPENDING ATTENTION

The Net seizes our attention only to scatter it.

This, says Nicholas Carr, is the paradox of the Internet that promises the greatest long-term influence over how we think.

Think of how intently you see people focusing on the Internet or smartphones (that businessman frowning at his phone, that teenager who can't hear you when they're at their computer).

The Internet seizes our attention. But then it scatters it because of all the different stimuli and messages it fires at us.

If attention were money, we'd soon stop and think. It's reckless spending. We freely deal out this currency of attention we don't value.

We spend a piece here, a piece there, a third piece on something we are not even interested in.

We watch a film on YouTube while checking an email that has just popped up, and Internet shopping. We answer a call on Skype and check our RSS feed while talking. We Google a question, and end up on a site via a string of links about an unrelated topic. We head to Twitter, and scan some tweets, while IMing and emailing at the same time.

Exhausted? You bet. But we keep on, and on, operating on what is termed shared attention, split attention, or continuous partial attention.

One in three of us feels overwhelmed, and is taking steps to limit our usage.[8]

But it's difficult when we get interrupted around four times an hour. Some 40 per cent of the time we never get back to what we were originally doing. The more difficult the original task, the less likely we are to resume it.[9]

We can't seem to pay attention and keep flitting to digital distractions while we do any task, and we feel tetchy, dissatisfied, stretched.

We're getting bad at choosing where to place our attention – because our brains are so overloaded. Our judgement is poor.

DO YOU LIVE WITH AN ALIEN?

Were you an adult BG (Before Google?) Or were you born this millennium? Do you remember a time before mobiles, never mind smartphones? Or have they been part of your life ever since you were old enough to count?

Today's teenagers are the first to have grown up knowing only a connected world. Yet today's parents of teenagers were themselves teenagers in a world that somehow operated without smartphones.

Parents and teenagers have always felt that the other was from another planet. But right now, they really are.

Maybe, of course, the alien in your front room is not of a different generation to you.

It is not just teenagers who are hooked on technology.[10]

> Teens, college-aged adults, and even middle-aged and older adults are hooked on their connective technologies and need to know who texted and who posted, and they need to know NOW. LARRY ROSEN

In a survey of more than 750 teens and adults in 2011, Larry Rosen split the respondents into iGeneration (born 1990–1998), Net Generation (born 1980–1989), Generation X (born 1965–1979) and Baby Boomers (born earlier than 1965).

He found that more than half (56 per cent) of Net Generation check text messages ‘all the time’ (more often than every 15 minutes), as do almost half (49 per cent) of iGeneration and more than a third (34 per cent) of Generation X.

The interruptions are so frequent that there is little – if any – time left to just think.

Constant connectedness is a particular issue for teenagers. A report by the Kaiser Family Foundation showed that 8- to 18-year-olds in America use media for 7 hours and 38 minutes a day, which goes up to 10 hours and 45 minutes a day including overlap using multiple devices at once.[11]

And children certainly start learning early. A survey of 2,200 mothers in 11 countries found that 70 per cent of 2- to 5-year-olds were comfortable playing computer games but only 11 per cent could tie their shoelaces.[12] Any parent who has seen a toddler try to ‘scroll’ a book or a block of wood as if it were an iPad will certainly have questioned the impact technology is having so early.[13]

BUT WE CAN MULTITASK

Teenagers today would have us believe – defending this view vehemently – that they are excellent at multitasking.

They may argue, in fact, that they have evolved to do it.

Look into the bedroom of any preteen or teen and you will see at least six forms of media engaging their attention at the same time, says Rosen:[14]

> Our research shows that they are likely to have the TV on; have music coming from an iPod, CD player, or computer; have the Internet running with multiple windows showing one or two social networks; be IMing at least three or more friends; and either be talking on the phone or, more likely, having a rapid string of back-and-forth text messages.

But the widely held belief about the ability to multitask – among teenagers and adults alike – simply is not true.

This is one of the big myths of the Internet, and it's no surprise we want to believe it when it comes to productivity, because, as Sherry Turkle notes in her excellent book *Alone Together*:[15]

> Our networked devices encourage a new notion of time because they promise that one can layer more activities onto it.

More time? Yes, please, it's exactly what we need.

Of course we'll take the option to do seven things at once.

But, though we like to believe we can operate like this, it simply isn't true – we can't.

A study proving this was done by Clifford Nass and fellow researchers.[16] They were fully expecting to uncover the benefits of multitasking, but they found just the opposite.

The researchers tested two sets of college students – heavy media multitaskers and light multitaskers, giving them problem-solving tasks to do.

One of the most ironic findings? Multitaskers were particularly bad at multitasking. It is one of the only things where the more you do it, the worse you get at it.

WATCH OUT, DIGITAL DISTRACTION ABOUT

I'm sure you've seen the digitally distracted out and about on the street, veering around in front of you, as if drunk, as they walk staring down at their phone.

While on that level, it's just annoying, distracted digital behaviour is dangerous – especially while driving.

An expert on distracted driving, Professor David Strayer from the University of Utah, says you are four times more likely to be involved in an accident while talking on a mobile phone.[17] The danger is even greater while texting – you are eight times as likely to be involved in an accident than a non-distracted driver.

Distracted walking causes serious injuries, too, and more than 1,000 people were treated in emergency wards in US hospitals in 2011 for injuries they got while walking and using a mobile or other electronic device.[18]

Walking into a lamppost, falling into a ditch and stepping out into oncoming traffic were just some of the accidents reported.

THE WORLD NEEDS 'AH-HA' PEOPLE

By being always available, constantly checking, you could be said to be switched on.

But, constantly connected, you are constantly thinking light, zooming, but staying nowhere.

The alternative to digital distraction is deep thought and reflection, where your attention is fully present.

These are the 'ah-ha' people.

The world needs 'ah-ha' moments and 'ah-ha' people who are not too busy to think. It needs them for development, solving problems, and innovation, to be creators, influencers, and clear thinkers.

What we've got at the minute is lots of scatterbrains.

There are plenty of people out there who have proved they can do the grazing – checking their email every five minutes, thoughts never fully formed. But, acting like this, we've forgotten how to focus on just one thing.

We need to redefine what is urgent.

Urgent is having creative thoughts. Urgent is intelligence, invention, bright ideas; not checking your email.

Prioritising does not mean walking around with your phone in your hand, so anyone who wants to can get your attention in an instant.

Since when did you agree to be 'on call'?

We are spending most of the time with only a fraction of our minds in any one place, rather than focusing.

We are not setting our own agenda. Our thought pattern is being led along a certain path by whatever demands are made of us by whoever contacts us in the places we automatically go, like email, Twitter and Facebook.

There is little room here for original thought. We're missing out on spells of genius, on lightbulb moments.

We need to reclaim 'ah-ha' time.

ON YOUR MARKS

Fun isn't being hunched over your smartphone at dinner. Or rising panic as you open your inbox before you're even out of bed.

It's time to pay attention to this.

It sounds simple. But it's not as simple as it once was, because we've got new distractions and our new habits have become ingrained. We feel like we can't leave the house without our phones (which we let constantly interrupt us), and we can't contemplate switching off our email. Some 70 per cent of us feel stressed if we can't get online.[19]

'We're currently the servant to the technology,' says psychologist Professor Cary Cooper of Lancaster University.[20] 'We have to take over and become the master.'

Perhaps you've got goals (life, work or personal) that you want to focus on achieving.

Yet the only thing that gets your attention these days is the 'pling' of a new message. There never seems time for the big, important stuff.

Split-attention and digital distraction are damaging our productivity – with massive implications for individuals and businesses. Digital distraction is also damaging our health and relationships.

Breaking this distraction cycle, and gaining focus, is what this book is all about.

Think how nice it is when someone listens to you properly, and you feel like you are being understood and valued. Now think how often you only pay partial attention to someone while you're texting ('hmmm?'), emailing ('what was that?') or on Facebook ('Yes, of course I'm listening').

In the next chapter we'll look at the skills we're losing, including listening.

Ready to aim for a world with a little less digital distraction? Great, then let's go.

A QUICK RECAP

- In one decade the whole digital landscape changed. Now we need to find a way to live well in it.

- New pathways have formed in our brains. We have literally rewired our mental circuits by using the Internet.
- Many people check their phones more than every 15 minutes for new messages. So many interruptions leave little time for thinking.
- We spend our attention freely and believe we can multitask. Evidence shows we can't.
- Digital distraction is dangerous while driving or even walking.
- The world needs 'ah-ha' people who have original thoughts. Make sure you don't rule yourself out.

FURTHER READING

- Nicholas Carr's *The Shallows* is a book at the centre of this debate. Carr looks in depth at what is happening to the brain. If you've only got time for one chapter, then go for chapter seven: 'The juggler's brain'. Carr, N. (2010) *The Shallows: How the Internet is Changing the Way We Read, Think and Remember*. London: Atlantic Books.
- Look at the table in chapter one of Larry D. Rosen's *iDisorder* to see how your digital habits compare to others of your 'generation'. Rosen, L. D. (2012) *iDisorder: Understanding Our Obsession With Technology and Overcoming Its Hold On Us*. New York: Palgrave Macmillan.

CHAPTER 2

The things we've lost

When is downtime, when is stillness? SHERRY TURKLE

WHILE WE LET OUR ATTENTION jump – dictated – from one digital distraction to the next, something gives.

We haven't given a thought to the skills slipping away. But these skills are many and vital; things like reading, remembering and even sleeping.

As we forge new neural pathways in the brain, we are often using old pathways less and less. For example, we're so busy checking Twitter that we let our interest in painting go by the wayside.

It is one thing when you lose the knack for speaking Italian because you haven't been to Italy for a while.

It's quite another when you start forgetting how to do once everyday tasks.

These are some of the things we are losing through digital distraction:

- Reading
- Solitude
- Memory

- Sleep
- Journeying
- Creativity
- Listening
- Learning
- Relationships

READING

Do you find it difficult to read a book these days? Do you struggle even to get to the end of an article?

The way we read online is very different from the way we read books.

We don't read on the web. We scan, as extensive research by web usability expert Jakob Nielsen has shown.[1]

You look at the homepage of one site for a few seconds, your attention skips to something else, you click on a link, scan a few lines, click away thinking of something else you want to do...

Does this sound familiar?

We are not just analysing text. We are deciding whether to click on links, navigating, searching and finding information, and processing pictures, videos and sound.

This experience is very different from the in-depth, long-form reading we used to do.

It could more accurately be called scanning or searching.

When we read on the Internet, says Nicholas Carr, 'Our ability to make the rich mental connections that form when we read deeply and without distraction remains largely disengaged.'

He talks about deep reading as a kind of deep thinking. He talks of the mind of an experienced book reader as 'a calm mind, not a buzzing one'.[2]

With a book, we go at our own pace, imagine the characters and the scene, and become deeply attached to the story.

In return for the time we put in, we get relaxation, pleasure, and fuel for the mind.

Another key difference with a book is that it ends (even if we have been reading it for weeks). The Internet is never ending. Nothing marks time there, which is one reason so much time is lost to it.[3]

Reading habits have also changed with the increase in use of ereaders. Some 41 per cent of people claim to have read more since getting an ereader.[4]

Do you read differently these days?

Have you stopped reading so many books, or do you read more?

SOLITUDE

Now, instead of solitude, we opt for broadcasting.

It's as if a thought needs to be posted as a status update to make it valid.

There's no need to be alone if we've always got a screen in our hand, a lifeline to everyone we know...

Why should we think for ourselves, or sit just with our thoughts?

Author and relationships expert Suzi Godson is afraid of the unhealthy psychological dependencies communication technology is creating for children:[5]

> Against a background of random beeping and pinging, how will our kids learn that silence creates space for creativity, that solitude can be joyful rather than lonely, that commitment does not necessitate constant connection?

We don't have to learn to rely on ourselves while our digital devices are there.

As soon as things get tricky (for example, we're bored of our own company) then – ping! – we can hop online and never have to face up to that difficult thing.

We're instantly diverted, instantly amused.

So we don't sit in solitude.

It's like a digital comfort blanket – but most of us aren't children any more.

> *Try sitting in solitude. Do you feel the urge to go and get your phone or laptop?*

Being alone can be challenging and boring. But it can also provide us with creative space, new thoughts, and a sense of calm.

As we increasingly define ourselves by what we say and do online (everything is a photo opportunity, not an experience), we are forgetting about the 'other us'. We spend so much time thinking about our persona that we have forgotten about the 'me' that can exist alone.

Sometimes experiences and thoughts are more powerful if they are not diluted by being shared with the whole www.

MEMORY

> The key to memory consolidation, or depth of understanding, is attentiveness; the opposite of what

> happens when we text, chat, surf and tweet at the same time. HELEN KIRWAN-TAYLOR[6]

If nothing gets our full attention, does memory suffer? Yes.

How might this look?

Two men go to a football match: one watches the match attentively, cheering with the crowd, tuned in to the atmosphere around him, commenting out loud on what he sees. The other keeps checking the Internet for the latest information about other matches, tweeting about the match, and texting people who are not there about what is happening. The next day, the man who paid full attention to the match has a vivid memory of it. The other man has only the status updates he posted.

What is happening in the brain here?

Memory is made up of two different sections in the brain – long-term memory and working memory.

Long-term memory is where we store all that we know. Working memory is what is happening to us right now. It holds a limited amount of information.

To create memories, things that are happening to us in the present are related to what we have known, and filed away in the relevant area (of similar experiences) in our long-term memory. To understand a new thing, we call on what we already know, pulling up information from long-term memory into working memory.

Nicholas Carr points out that we are 'flooding our working memory with information' when navigating the web.[7]

We are so busy doing so many things there that all we can do is process on a surface level. Carr says that the depth of our intelligence depends on how well we can transfer information from working memory to long-term memory. Without this: 'Our ability to learn suffers, and our understanding remains shallow.'

If working memory is full, both recall and the laying down of new long-term memories can't happen as they are meant to.

Mind expert James Borg explains that there are three elements of memory – encoding (taking in information), storage, and retrieval (recalling information).

Memory fails if there is a breakdown in any of these elements. Borg says the reason for most memory deficiency lies at the encoding stage – because we don't pay attention to things:[8]

> If we don't focus on what we're listening to or what we're seeing then the experience passes us by. Quite often it's because we may be trying to 'process' more than one thing at a time.

If there is no encoding because of split focus (you're tweeting and texting and IMing while you watch the match) then there is literally nothing to store – nothing has gone in to your long-term memory.

Another recent change is that we are treating the Internet as an external memory. If we expect we will be able to access information later, we are more likely to remember how to get to it, and less likely to remember the information itself. Betsy Sparrow calls this 'The Google effect'.[9] This is an example of a transactive memory system, where information is stored collectively outside our own mind. Doing this is not new. We use other people to store information, too (one person in a couple will remember when everyone's birthday is, for example).

But putting nothing into our own memory bank could have a long-term impact. The memories we form can enrich our whole future, as we flash back to tangible moments from our past.

We'll look at this topic again when we talk about switch-tasking in Chapter 8.

SLEEP

Have you noticed any difference in your sleeping patterns since you started using digital media more?

Do you ever dream, for example, about tweets?

It may sound silly, but if we are connected to our phones right up until the moment we go to sleep, no wonder digital distractions are crossing over from RL (real life) to the land of nod.

More than 80 per cent of smartphone users leave their phone switched on all of the time – even when they are asleep. Half of these users admit to using their phone after it wakes them up.[10]

That means text messages are being allowed to elbow out vivid dreams or deep sleep.

Which is more important?

Even if your phone is switched to silent, a part of your mind is aware messages might be arriving. Have you ever got up in the middle of the night and checked for messages while you were up?

Checking messages and emails before you are even out of bed in the morning means other people get to dictate your thoughts before the sun is up.

And checking them just before you go to bed not only means a chance they will determine your dreams, but stops your body preparing properly for sleep.

Light-emitting devices – like smartphones or computers – inhibit the body's making and releasing of melatonin (the hormone produced in response to darkness, telling the body to prepare for sleep).[11]

One simple trick could improve your sleep – turn your phone off at night, and turn your technology off earlier.

Think about your sleep habits by answering these questions:

1 What time do you disconnect from:
   - smartphone or mobile
   - Internet
   - emails
   - social media?
2 Do you ever get interrupted by your mobile/laptop while sleeping?
3 Are you often tired?

JOURNEYING

I've started having to bite my tongue on trains and buses. Everywhere, there are people staring down at their smartphones.

What about the world going by outside?

We used to think of journeying as classic thinking time, but now we see the chance and grab it (five whole minutes!!!) to shoehorn in yet another bit of digital processing, and take another hit of information overload.

We are doing this at precisely the point where we could give our brains a much-needed and welcome chance for downtime, for wandering.

Inspiration doesn't generally arrive while you're checking your emails.

Have you ever had an epiphany moment on a journey?

Do you let yourself gaze out of the window and just think?

On journeys, as well as in the general run of day-to-day, we've started stopping thoughts half-way down their track.

We'll be thinking something, and then 'beep beep' or 'ping' and off we go, our almost-formed thought left discarded. Somewhere there is a room of lost thoughts and half sentences that has been getting far too full.

Thoughts – and discussions – need to go to one place, then the next place, then the next place, then the next place... and finally on to an idea. It isn't an express route. Sometimes it takes a lot of turnings to get there.

But if we let our thinking be cut off mid-flow we are not even giving our ideas a chance to form.

Another thing in danger of being left discarded at the side of the road is discovery. Before, you'd arrive in an area, looking for somewhere for dinner, and wander round the streets, noticing things, while deciding on something that looked good.

Now the discovery is confined to Google. The decision is based on which restaurant website shows up in the search rankings. We might do an extra bit of meandering by reading some customer reviews. When we get geographically nearby, we'll be head down, navigating on our smartphones. What was the local area like? We really haven't a clue.

CREATIVITY

> The important thing is to create. Nothing else matters; creation is all. PABLO PICASSO

Creativity might be something we think applies only if we are pursuing something such as writing, painting or design. But it goes far beyond that.

Look around you at work, at home, and in the environment you live in – where can you see creativity?

We need it to have great ideas, to change things, and to dream, as well as to create art in all its forms.

But what happens when you are creating and then go and check your email, or have a little look at the Internet?

Productivity expert Leo Babauta says distraction hurts creativity.[12] He makes a distinction between creating versus communicating and consuming, and says the two cannot be done at once:

> We lose a little bit of our creative time, a little bit of our creative attention, each time we switch. Our mind must switch between modes, and that takes time. As a result, our creative processes are slowed and hurt, just a little, each time we switch.

Babauta suggests separating your day into a time for creating and a time for communicating and consuming – but never doing the two at once.

Communicating and consuming can be useful – they provide a chance for collaboration and inspiration, he says, but we should be wary of the time they take that could be used for creating.

Have you ever been frustrated by digital distractions when you are trying to create?

Next time could you block off time just for creating?

We need to be wary of being tempted by distractions when things get difficult. We can be quick to log in, grab the comfort blanket of digital distractions, and abandon whatever it was that was tricky or challenging about the creative process.

LISTENING

You can tell if someone is not listening to you while they're head down, checking their phone, 'mm hm'.

Author Susan Maushart describes it as 'a hooded quality to their voice that's kind of the vocal equivalent of a blank stare or a busy signal'.[13]

Many people – particularly teenagers – now don't expect 100 per cent attention. They expect that their friends will be IMing at the same time as texting, doing their homework, and browsing Facebook.

What does this say, when giving full attention is seen as too much to expect?

Really listening – in a time when we face so many distractions – is one of the quickest and easiest ways to reconnect to the people around you.

The messages you are giving are: I'm not rushing off to check my phone. Let's talk. What do you think? I'm here.

It strengthens your bond with the person you are listening to. You also remember the conversation.

LEARNING

Many teachers and lecturers claim that pupils are digitally distracted in classrooms and lecture theatres these days. I heard a story of one lecturer who asked an (almost 20-year-old) student to turn off their phone during the class, to which the student replied, 'But I've never switched it off.'

The brain has to work hard when you learn and, if you're digitally distracted, you make learning harder still.

The way young minds are working needs to be understood. As well as how to teach things, we need to think about what needs to be taught.

Brain expert Susan Greenfield says we need to understand what is happening to the brain, so we can plan education:[14]

> Only once we have appreciated how our malleable brains are interacting with the current and imminent technologies will we be able to plan the kind of education, as well as the goods and services, which we shall both want and need by the mid-twenty-first century.

She points out that ways of thinking have changed – such as thinking of things with menus and fixed options as they would be online, rather than thinking laterally.

Some experts say attention skills should be taught in schools – while children's brain networks are still at a sensitive stage of development.[15]

Do you switch your smartphone off while you're learning?

RELATIONSHIPS

Have you noticed that people have stopped phoning each other for a chat?

A distance is being put between relationships that didn't used to be there. We haven't got time, for a start... and it just seems, well, easier somehow, to email.

> The flight to email begins as a 'solution' to fatigue. It ends with people having a hard time summoning themselves for a telephone call, and certainly not for 'people in person'. SHERRY TURKLE[16]

In the UK, day-to-day, people now text more than they call or talk face-to-face.[17] In America, 50 per cent of people prefer to communicate digitally rather than in person.[18]

With such a never-ending stream of information coming at us, no wonder we feel exhausted from processing it all, and find it difficult to summon up the energy for meaningful communication.

We send, and process, so many messages in a day, that having the energy to be original or writing a message with feeling seems too much and becomes a rarity.

> *But what if you sent just three messages that meant something today, rather than 30 messages without much meaning?*

Often the solution we turn to is broadcasting – via social media. Send one message to a bunch of people and we can tick that off the list – communication done.

This is where much of our time is spent. But do we even know these people very well? Perhaps not.

Research by Cameron Marlow shows that the average Facebook user with 120 friends will use two-way communication (such as emailing or chatting) with only five of them – that's just over 4 per cent. Users with 500 friends will communicate (directly interact) with just 10 of them (if they are a man) or 16 of them (if they are a woman).[19]

These new relationships are something we will look at later on in the book. We'll also look at the many tensions digital distractions are causing in relationships.

A QUICK RECAP

- On the web we scan and search, rather than read and think deeply.
- Solitude is a forgotten pastime in an age where we constantly have a mobile device next to us.
- If attention is divided, memories are not encoded, and there is nothing to recall.
- Are we giving up on the luxury of sleep by leaving our smartphones switched on?
- Do you use journeying time to take in more digital information, or to think?
- Don't try and create at the same time as consuming or communicating; you'll sap your creativity.
- Giving 100 per cent attention to the person you are talking to is a way of listening that is being neglected.
- We need to think about the way we learn, and what we need to learn.
- Relationships have changed dramatically in the digital age, and we are less likely to communicate face-to-face.

FURTHER READING

- Research by Jakob Nielsen underpins much web usability work, particularly to do with how we read on the web: Nielsen, J. http://www.useit.com/.

- James Borg's *Mind Power* has an excellent description of how memory and attention are linked, as well as great ideas about the power of thought (Chapter 7): Borg, J. (2010) *Mind Power: Change Your Thinking, Change Your Life*. Harlow: Pearson Education Limited.

CHAPTER 3

Mind pollution – the four digital evils

We live in curious times. It's called the Age of Information, but in another light it can be called the Age of Distraction. LEO BABAUTA

FOUR DIGITAL DISTRACTIONS THAT ARE particularly hard to ignore are email, smartphones, social media and the Internet.

The reasons we turn to these digital platforms are complex.

These are some of the reasons we use them:

- Boredom
- Ego
- Social connection
- Communication
- Seeking social validation
- Learning
- Fun

- Fear
- Need to be needed
- Knowledge/information
- Habit
- Addiction
- Distraction (which breeds further distraction).

Which of these reasons do you identify with?

When it comes to distraction, each medium has its own particular pull. We'll look at the Internet, email, mobiles and social media in turn.

I'LL JUST CLICK HERE

> I write a sentence, maybe two, and then I LOOK AT THE INTERNET. I write a few ideas down, and then I LOOK AT THE INTERNET. Before I even begin, I LOOK AT THE INTERNET. DAVID BADDIEL

This description by David Baddiel, about how he has never written a column in one go, observes brilliantly a problem many people will identify with.[1]

Even once he has begun, the Internet lures Baddiel back again:

> I get through half a paragraph and something in me thinks, often reflexively now (my hand has clicked the mouse before I've even registered it), 'Phew: that's enough toiling away at the coal face – time for a long, cool drink of YouTube/Facebook/Xvideos…'

So that's where all our hours go...

We switch on our computer, and our hand reaches automatically to open the Internet. We plan to spend five minutes there, tops. Then, somehow, the morning has gone.

One great paradox of the Internet is it gives us the illusion of going fast. We whiz from site to site. We don't even read, we scan.[2] We are highly impatient; and we don't wait around if we can't see what we want. We skip off as soon as an interesting link catches our eye. There's a whole enticing web to navigate, from link to link to link. It makes us feel oh-so-knowledgeable... we love the way so much information is at our fingertips. But by the end of another day of this, we're utterly drained.

One huge time-draining issue with it is that is has no timer on it, and no beginning or end. So we often end up spending longer than expected there. It's so useful, there are so many possibilities, that we always seem to end up doing just one more thing...

Pre-smartphones, we accessed the Internet roughly five times a day in longer chunks. Now we access it 27 times a day.[3]

Do you ever use the Internet as an add-on to whatever you're doing? Always there, almost unnoticed in the background, like the paint on the walls?

Then, whatever else you're doing, you'll just add the Internet in... so you might watch a film while Internet shopping. Or ask Google to prove a point in your conversation.

What's wrong with this approach?

The shopping takes longer, and you don't absorb and remember the film. You have to read then re-read the information when you lose your place on the shopping website. You are more likely to make mistakes.

We think we can do two things at once, but we can't. We'll look at what's happening to our attention here as we go through the book.

As we cruise around the Internet, our attention is what everyone wants. They try different ways to grab it: flashing, flickering lights and pictures, funny one-liners, special offers...

So far, we've given our attention away fairly freely. But that's about to stop as we realise how valuable and limited our attention is, and begin to carefully and purposefully spend our daily quota.

Many of us feel like we're at the point of information overload. We'll look at focus methods throughout Part 2 that will help you deal with this.

On one level, we've saved space, we've stopped storing information. Got a question? Just ask Google. We've stopped remembering things, but we've developed new skills – of searching and of spotting information quickly.

Overall, we're processing far more information than we did. We consume three times as much information as we did in 1960.[4] And the amount of data we create as a society is staggering. Predictions suggest that by 2025 the amount of data stored is expected to reach 100 zet – the equivalent of 36 billion years of HD video.[5]

While your personal Internet use is but a pebble in the ocean of information, think about how your consumption adds up, taking up space in your mind.

Notice the next time you click automatically and find yourself online without even realising you were going there.

How long did you stay? Longer than intended?

WHO NEEDS EMAIL?

Other people; don't you just love 'em?

They cc you in, bcc you in, forward you emails, and forward your emails to other people, reply to all, and send you long rambling messages... until your inbox is groaning at the seams.

Corporate email storage is growing at 20 to 25 per cent a year, and an estimated 30 per cent of the traffic is occupational spam – caused by overuse of cc, bcc and reply all.[6]

So, most of this email, we don't even need.

One company that has had enough and decided to ban email is Atos. They found that employees thought only 15 per cent of their messages were useful – meaning processing the rest was lost time.[7] We'll look at their story in depth in Chapter 6.

Most businesses haven't been very imaginative, yet, when it comes to email, so far going for a one-size-fits-all always-on approach.

What about you? Do you think your email is a waste of time?

How much do you use it?

Let's do a quick analysis. Here are a few questions:

Email usage quick analysis

- Do you have your email open constantly (either on a screen or mobile device)?
- How many times a day do you check your email?
- What time of day do you check your email?
- How many emails do you send a day (look at your sent mail to work this out)?

- How many emails do you receive a day on average?
- How often do you completely switch off your email for more than 24 hours?
- Do you ever check your email when with others – e.g. at the dinner table?

Got your answers? Great.

Were there any surprises in there?

Let's zoom in on just one of those questions: How many times a day do you check your email?

What's your answer?

Focus expert, Jurgen Wolff, has some no-nonsense advice here: 'Set a number of times each day that you will check your email. Do you have a number in mind? Good, now cut it in half.'

So what's your answer cut in half?

Wolff says there are very few people who need to check more than four times a day, and that two or three times a day is even better.[8]

It's easy to blame other people for our overflowing inboxes. But part of the reason we check email so often is that we like the buzz. Each message potentially gives us a mini-reward. It could be urgent. It could be an opportunity. So we keep checking… even though most of it isn't.

But these rewards come at a delayed cost. Our email mountain has become so huge and we're overwhelmed, sometimes even declaring 'email bankruptcy'.[9]

John Naughton (journalist and writer) says email has 'become a monster that's destroying our lives', infiltrating leisure time, family time, and even sleep time.

With the arrival of the smartphone, he says, there came an ever-lengthening of the working day because of an expectation that people will be always contactable by email.[10] Following on from this is the expectation that any message will get a quick response.

Delayed response

But what happens if you don't respond immediately?

Perhaps not quite what you'd expect.

Email begets email and, as you start to slow down your email checking, fewer emails start coming in.

There is often not the need we think there is or say there is for 'now, now, now'.

People get used to our email policies. They start respecting our time. We start using email more sensibly and our email load suddenly drops.

We may act out of fear (of a boss or a client) or to be seen in a certain way (as working hard or available, for example). This may mean feeling a pressure to keep email switched on.

To turn round in such a climate and say you are not checking your emails takes a lot of confidence and courage. But would your boss rather you get to lunchtime having done (a) nothing but process emails, or (b) your work?

Priority number one

Email is a huge source of distraction and frequently knocks us off course from what we were doing.

Why? Our priority for the day is often shoved to one side – never to be picked up again – as we re-prioritise to deal with other people's demands. We even log out of email realising that we didn't even send the email we went to 'just do' – because we got so wrapped up in what other people wanted.

Productivity expert Leo Babauta says that you should not check your email first thing in the morning. This is something most people find extremely difficult, but it is a really important, simple tip.[11]

If you do check email first thing, Babauta says, you are in danger of getting stuck in your inbox and not getting out of it.

We often feel anxiety and a need to respond.

> This need to respond gives us anxiety until we've responded, but unfortunately, there is a never-ending stream of things that require your response.
>
> If we allow these messages to force us to respond, almost as soon as they come, then we become driven by the need to respond. Our day becomes responsive rather than driven by conscious choices. We flit from one task to another, one response to another, living a life driven by the needs of others, instead of what we need, what we feel is important. LEO BABAUTA[12]

It's only a little email, yes. But it takes your time, your energy and your thought to reply to each person. If you check it constantly, it uses up a large portion of your precious attention.

Go to zero

The system I use on email is Inbox Zero, invented by Merlin Mann.[13]

Every time you check your email, you clear your inbox to empty. Then you log out, with no email 'To Do list' on your mind. You limit the number of times you check email, and you limit your time in email. You soon become very aware of any emails that are wasting your time.

Say there was a limit, and your email account only allowed you to send three emails per day. Which three would you send?

What if you never had to check your email again? What would you do with that time?

Your answer gives a clue – as to what your email could be stopping you from doing.

LIFE IN A LITTLE BLACK BOX

It wasn't so long ago – relatively – that we were learning how to switch our phones on (do you remember your first mobile?). Now, it seems, we have forgotten where the off button is.

In the UK, nine in 10 people own a mobile (compared with only 36 per cent in 2000). More than a quarter of adults and nearly half of all teenagers own a smartphone.

Ask these smartphone owners if they are 'highly addicted' to them, and 37 per cent of adults and 60 per cent of teenagers will say yes.[14]

The problem is, these phones access all areas, and we have drawn no lines.

Text messages shove aside dreams. Calls are taken on holiday. And the weekend starts to take on many characteristics of the working week.

Before breakfast many of us chomp through a list of updates or stomach demand after email demand.[15]

Things often start out innocently enough. The smartphone arrives as a company 'gift', perhaps.

But then we adopt compulsive checking behaviour.

Price compares users of phones such as a BlackBerry with those of older phones. The BlackBerry (or equivalent) users had only 13 per cent more emails, but spent two and a half times as long as the others checking email outside work hours.[16]

'It is as if simply owning a BlackBerry, or similar device, drives this compulsive checking behaviour,' he says.

We used to use the term 'CrackBerry' a lot. We've gone quiet about that now. Could that be because such a level of addiction has become almost the norm?

Emergency on planet brain

Our mobiles and smartphones expertly yank our attention from wherever it was, distracting us entirely.

It beeps! There's a flashing light! Emergency on planet brain!

If there's a new message, we can't resist checking. This is a pleasure-seeking response, the message gives us validation, and props up our egos – someone loves me! A message! I must see it NOW!

If we leave our phones on, an interruption could come at any point.

In the UK, we send 129 billion texts per year.[17] These interruptions we face are very frequent – as we saw in Chapter 1, many of us check for messages more than every 15 minutes.

When it pings, or beeps, your mobile physically pulls you away from what you were doing. You deal with the message, and maybe then become distracted by something else. Twenty minutes later you are finally back doing what you were doing. So the distraction cycle goes.

Joe Kraus, a partner at Google Ventures, says for teenagers, interruptions are even more frequent:

> Do you know what the average number of text messages a 13–17-year-old teenage girl sends and receives every month? The average? 4,000. That's one every six minutes that she's awake. Boys aren't much better at 3,000.
>
> Think about that. You're interrupted once every seven minutes.
>
> What kind of culture is that creating? What kind of mind training is that doing?

We are over-developing 'quick thinking, distractible' parts of the brain, he says, and letting 'long-form-thinking, creative, contemplative, solitude-seeking, thought-consolidating pieces' of the brain waste away by not using them.[18]

Even when our phones are not pinging or delivering messages, we are on a kind of standby, waiting for them to. And out of sight doesn't mean out of mind.[19] Even if our phones are in the next room, we keep a bit of our attention focused on them. So we don't concentrate as powerfully as we could on the task we are doing.

As soon as I realised this, I switched my phone off while doing tasks I wanted to concentrate on, like writing. Only once you switch your phone off do you realise how much you were 'just checking' before.

Lifelines

Have you ever thought you heard your phone beep or vibrate when it hadn't? Have you ever patted your pocket or looked in your handbag to 'just check' your phone is still there?

'Phantom ringing', where you think you heard your phone ring or buzz, is just a small indication of how attached we've become.

'Phone-in-hand syndrome' – people who walk around with their phones in their hand at all times – is another.

Our phones almost seem to have morphed into an extension of our being.

'Are they welded to these things?' someone said to me, recently. 'It's as if it is part of his body' a parent of a 22-year-old commented.

Unhealthy attachment?

Many of us get distressed if we are separated from our phones for any length of time.

Rosen talks about a conference where the speaker asked 500 people to give their phone to the person to their right to put away. Some 15 minutes later the speaker asked how they were feeling and most of them said they were so distressed that they couldn't focus on the talk.[20]

> *Do you feel anxious when you can't check your mobile device when you want to?*

Rosen offers a note of warning:

> Anxiety-related problems, such as OCD, seem quite likely to develop, given the attachment we have to our technology and technological devices.

Here are three questions to think about:

1. *What boundaries do you have, or would you like to set, for your mobile or smartphone?*

2 *Where should it not be switched on?*

3 *When should it not be switched on?*

SOCIAL MEDIA

For many people, social media are particularly addictive.

Social media draw us in and gobble up our time precisely because they are a social tool.

Tied up with relationships and the construction of our own identity, the pressure and the stakes are high. We want to appear a certain way. We agonise over our profiles. We wonder what to say and how to appear. All this takes attention and time.

We are also eager to get insight into other people's lives. There's a window there for us to look through any time we want, so we spend hour after hour staring through it... even if we don't know the people.

Time spent carefully constructing and maintaining our online identity distracts us from spending time on elements of ourselves we might want to develop not just for the sake of broadcasting.

There is a danger we forget about doing things because we want to, while we focus and spend time on doing things for an audience.

Don't miss a minute

One thing that keeps us hooked on social media is what MTV terms FOMO – fear of missing out. FOMO keeps us checking in all the time – just in case something has happened. In a survey of young people, 58 per cent said that when they were unplugged they worried they were missing out on something.[21]

While we're not happy when we're not there, we're not necessarily happy when we're there either. Everyone else seems to have such a wonderful life. But where is our wonderful life?

'Reading about your friends' wonderful lives leads to the erroneous conclusion that everyone is happy except you,' says Clifford Nass, a professor of communications at Stanford University.[22]

But when posting on social media yourself, you are likely to talk things up. So it sounds to everyone else like you have the wonderful life.

As one friend I talked to said, it's a 'peacock tool'.

Next time you are looking at status updates of other people, try 'translating' them down to reality.

Making friends

One big thing distracting us on social media is our friends. There are so many of them, for a start...

Social media have changed what we call a friend. But some of the hundreds of 'friends' taking up our time might not actually be the people we most want to spend our energy on. We read their status updates, we look at their photos, but in fact we hardly know them. Meanwhile, our old friendships lie off to the side, discarded.

Adults and teenagers are spending so much time on 'screen-life' friendships, it leaves little to invest in 'real-life' friendships, author Sherry Turkle points out.[23] We only have so much time and attention to use in a day.

Our inclination is becoming more and more to keep things on screen – we're getting out of the habit of meeting up or talking on the phone. We don't even want to.

We settle for being 'Alone Together' (the title of Turkle's book). We feel we have caught up when we have checked Facebook, looked at a few photos, and sent some emails. We send out a 'broadcast' to lots of people, rather than communicating many times over one-to-one (a further snub to the friends we used to confide in, who read on our blogs stories that used to be just for them).

Turkle talks about intimacy, about being with people in person, hearing their voices and seeing their faces, 'trying to know their hearts'. This is what we have lost.

Here are two questions to think about:

How much do you conduct friendships online?

Which social media platform is the biggest time drain for you?

A QUICK RECAP

- Time tends to slip away from us on the Internet as we take a never-ending journey we didn't even plan.
- Do you use the Internet as a backdrop to whatever else you are doing, so you end up multitasking?
- How often do you check your email?
- Bad email habits are wasting time for businesses and individuals.
- When you check your email, you often get distracted by demands made by other people and forget about what you wanted to get done.
- For such a small device, the smartphone demands a lot of attention.

- How often do you get interrupted by a text message arriving? Do you ever turn your phone off to stop interruptions?
- Do you have boundaries for smartphone or mobile use? Think about what these could be.
- Creating and maintaining social identities takes a lot of time, and has high stakes.
- Fear of missing out (FOMO) is a big pull on social media.
- We could be neglecting face-to-face friendships while we spend our time instead on online friendships.

FURTHER READING

- A key book in this debate is *Alone Together* by Sherry Turkle. I highly recommend it. It is full of colour and insight. Turkle, S. (2011) *Alone Together: Why We Expect More From Technology and Less From Each Other*. New York: Basic Books.
- See http://www.merlinmann.com/ and http://inboxzero.com/articles/ for more information about Inbox Zero.

PART 2

Attention

---

noun the activity of attending, esp. through application of the mind, to an object of sense or thought

CHAPTER 4

Assess

I predict that spending so much time in cyberspace will inevitably lead to minds very different from any others in human history. SUSAN GREENFIELD

HOW DIGITALLY DISTRACTED ARE YOU? Digital distraction occurs on a spectrum – from non-digital users who are not distracted at all, to users who have severe Internet addiction.

The danger is that we focus on Internet addiction, thinking of it as a severe problem, and don't notice that a milder form we ourselves may be suffering also has serious consequences.

Our behaviour does not need to be defined as full-blown addiction to be damaging.

Just because we don't meet the criteria for addiction doesn't mean aspects of our psychology are not affected. And it doesn't mean we are immune to it.[1]

> Even if we do not meet the definitions for 'Internet addiction' and because of that are not captured in the statistics, our psychology is still affected by the Internet, and how it is affected deserves to be investigated and understood. ELIAS ABOUJAOUDE

Wherever you are on the digital distraction spectrum, in a few weeks, you should be able to say that you are less digitally distracted.

Those whose attention is currently ruled by digital distraction will find this happening less. Those who already have digital balance will find new ways to maximise deep thinking.

Let's get started and assess your level of digital distraction.

ASSESSING YOUR DIGITAL DISTRACTION

We can guess at how much we are using digital devices and platforms like social media and email but, in doing so, we would probably underestimate our usage.

Let's go back to the pile of digital distractions you gathered – mobiles, tablets, laptops, smartphones, computers, iPads, iPhones and BlackBerries – to build a bit of detail into a picture of your digital usage.

First specimen under the microscope – your phone.

Take a look at your mobile or smartphone.

- How often do you switch it off? Ever?
- How often does it 'call out' to you with updates and messages and calls?
- Do you go straight to it when it does?
- How often do you find yourself hunched over it typing?

Now look at your computer, laptop or tablet.

- Which applications or programmes do you use on it the most?
- Which sites do you go to automatically?

- How often do you use it?
- What time of day do you switch it on? And off?

Now, thinking in general terms, ask yourself:

Are you pulled to one digital distraction particularly?

- Email
- Social media (including Facebook and Twitter)
- Smartphone
- Internet
- Other (name it).

Onto the scales

We have seen that digital distraction exists on a spectrum, from serious addiction (for which you should seek professional psychological help) to moderate or non-use.

If digital users were split into three groups, which group would you be in?

These are some of the traits of these three types (intended as a general guide only).

Sometimes, traits from one type may be experienced by a person of another type who has a weakness in a specific area (being constantly on email, for example).

See which group rings true the most for you.

Which type are you?

Dangerously distracted

- Has email constantly on when computer or phone is on.

- Rarely switches off smartphone (if ever).
- Responds instantly to texts, updates, emails, messages.
- Spends more of the day switched on and using digital devices than switched off.
- Frequently simultaneously uses many different platforms – 'multitasking'.
- Frequently experiences long stretches of time 'disappearing' while online.
- Carries phone in hand at all times.
- Rarely communicates face-to-face or by phone call.

Slippery slope

- Sometimes loses track of chunks of time while browsing the Internet.
- Often does two or more things at once (e.g. Internet shopping while emailing).
- Uses social media without screening out certain feeds or setting a time limit.
- Spends longer than wanted on email.
- Defers to phone in company or before other things that need doing.
- Often gets derailed from tasks by demands from other people via social media/email/phone.

Mostly focused

- Tends to use one platform at once (singletasking).

- Switches smartphone, mobile, laptop, tablet, etc. off on a regular basis.
- Doesn't respond instantly to messages.
- Has decided upon an email policy which is not default 'always on'.
- Is very aware of exactly how long is spent online for any digital task.
- Rarely gets distracted from a task.
- Has long spells of being disconnected.

If you're in one group and want to move up to a less distracted group, pinpoint some of the areas you want to start working on. Perhaps circle the things you most want to work on. These are some new behaviours to begin to aim for.

DIGITAL DISTRACTION DIARY

Now you've got some idea of your level of distraction in broad terms, and an idea of which digital platforms could be causing this, it's time to see exactly what is causing your distraction (whether that be mild, moderate or severe).

Many people are surprised at what a big influence their digital behaviour is having over their life and their work, once they start looking closely.

Get yourself a sheet of paper or a notebook and a pen, and write today's date on it. Done that? Great.

This is your distraction diary. It doesn't need to be anything fancy – the important thing is that you carry it around with you in your pocket or handbag all day.

Each time you are distracted, simply note down the time, what you were distracted by, and what you were distracted from.

Here is a template:

Time:	
Distracted by:	
Distracted from:	

For example:

Time:	11.30 am
Distracted by:	Text message
Distracted from:	Walking down street thinking

Time:	2.15 pm
Distracted by:	New email sent to me
Distracted from:	Work I was meant to be doing

Time:	7.30 pm
Distracted by:	Facebook
Distracted from:	Conversation in living room

Get the idea?

It's important not to judge, just observe it and note it down, then carry on.

What about if you intend, for example, to log in to email for 20 minutes?

That's fine – you can note it down if you want (4.07 pm: planned email session for 20 minutes) but you don't have to write it down.

Also note down any distraction that tempts you but that you resist.

For example:

Time:	4.47 pm
Tempted by:	Going off on Internet meander
Almost distracted from:	Writing mind-numbingly dull report
Resisted digital distraction by:	Getting cup of tea instead

Aim to keep your distraction diary for a whole week (it helps to see what you do on different days).

At the end of your week, clear a bit of time (20 minutes should be enough) and sit down and have a look at your distraction diary.

Get a pen or pencil and circle anything you notice.

Is there one thing that comes up again and again? Do distractions come at a certain interval? Do distractions always come at the same time of day?

Look for any patterns you can spot.

Remember, this information is just for you. It's to help capture data about your digital distraction habits.

Looking at your distraction diary for the week:

- What surprised you?
- What were you pleasantly surprised by?
- What seems to be your number one distraction trap?

Don't throw away your diary – we'll look at it again in the next chapter when we look at some common distraction patterns like trigger times of day and trigger tasks.

TIME FOR A TANTRUM

Are you tempted to put the book on the shelf and give up yet?

Working through the process of tackling digital distraction is particularly difficult to start with (remember all the automatic pathways in the brain you are fighting against), so you'd be perfectly entitled to a tantrum.

Go and get yourself a cup of tea or coffee as a well done for getting this far. There's no pretending that it's easy.

You might be particularly frustrated once you start to realise how much time and energy digital distractions are 'taking' from you, yet you haven't yet learnt to stop being distracted by them.

Acknowledge your frustrations by writing them down (*'It's too hard.' 'I'm getting nowhere'*, etc. etc.).

The good news is that, as you get towards the end of this process, you can expect things to change.

You'll feel calmer, less stressed, and healthier. You'll have a strong sense that you are spending your time better. You'll notice the alarming rate of progress you are making on projects that mean something to you.

If you're really frustrated at this point, skip straight over to the next chapter and start working on your focus…

In the rest of this chapter we'll look at the mind, and at a definition of Internet addiction.

MIND GAMES

Digital communication can make us feel a lot of things. It can make us feel angry, overwhelmed, stressed, panicked, fearful, inadequate, pressurised to do things, jealous, small, or left out. It can also make us feel empowered, knowledgeable, happy, interested, inspired, connected, and like we belong.

Sometimes our emotional responses are hard to unpick. There may be reasons why we stick with digital distractions that are not at first glance obvious. These might be, for example, to feel important, needed, or busy (so we don't have to face the gap that is there if we free up our time). They might be out of fear of missing out on something, or fear of another person.

Psychology plays a large part in digital distraction and we will look at this alongside productivity throughout Part 2.

A great book for more information on negative and positive thinking (as well as memory) is *Mind Power* by James Borg.[2]

Often the difference between feeling positive or negative is not in what we actually see, but in how we think about what we see.

A negative thought cycle might go like this:

'Urgh. So many emails. I don't even know where to start. A waste of my time. Read first message. Feel angry. Fire back instant angry response. Maybe shouldn't have sent that. Overwhelmed by all these emails, can feel stomach tightening at how many I have to get through. Boss will wonder what I've been doing. Will be angry when I don't hand report in. Will lose house if get fired. Will be homeless and poverty-stricken…'

A more positive response might be:

'Expected to check email and have to do work, too. Case of lots of demands on my time. So email can only take 15 minutes. Timing it from now... Delete spam without reading it. Read email from boss first and send quick reply so she knows I've seen it. That email from a client makes me angry. Will archive for later – their To Do list not mine and don't want to respond in heat of moment. Compose and send message I was planning to send. Quick message to friend about meeting this weekend. Log off email now and on with tasks for the day.'

Remember to check up on your thinking as well as the digital distractions you are facing. Being positive about things is often a way to be more powerful and productive.

THE SERIOUS SIDE

Internet addiction is a serious issue, and one that is increasingly being recognised as such.

> The Internet addiction diagnosis as most experts have defined it is concerned with two main features: a) too much time is wasted online (regardless of what it is wasted on), and b) the time taken from other activities and other relationships results in a major loss to the individual, such as divorce or significant difficulties at school or work. ELIAS ABOUJAOUDE

As noted earlier, this year, for the first time, Internet addiction will be included in the manual used worldwide to classify mental disorders – the *Diagnostic and Statistical Manual of Mental Disorders* (DSM). It will be marked 'for further study'.

Whether it should be included in the DSM has sparked much debate among psychologists, some of whom think that it is simply a new manifestation of disorders such as compulsive shopping or gambling.[3]

Internet addiction can be said to be similar to impulse control disorders, substance addiction and OCD and, as Aboujaoude states, it is along these lines that researchers have developed questionnaires for Internet addiction, using criteria for these conditions borrowed from the DSM.

Extreme examples of digital addiction show us what happens when things go too far. Some shocking cases have been revealed, such as one couple in South Korea addicted to the Internet who let their three-month-old baby starve to death while they raised a virtual baby online.

South Korean police arrested the 41-year-old man and 25-year-old woman, who reportedly left their baby at home for hours at a time while they went to an Internet café. They became obsessed with raising a virtual daughter called Anima in the game Prius online. Their real baby died from malnutrition.[4]

South Korea is one country which has taken action against widespread Internet addiction by setting up Internet addiction clinics, bootcamps and hospital programmes.[5]

Studies have documented Internet addiction in countries including Italy, Pakistan, Iran, Germany and the Czech Republic. Reports suggest it has become a serious public health issue in China, Korea and Taiwan.[6]

Dr Kimberly Young presented the first paper on Internet addiction in 1996 (entitled 'Internet Addiction: The Emergence of a New Clinical Disorder').

She has also developed a questionnaire to diagnose Internet addiction, called the 'Internet Addiction Diagnostic Questionnaire' (IADQ).

This Internet addiction test is reproduced from her book, *Caught in the Net*.[7]

To assess your level of addiction, answer the following questions using this scale. Please tick your chosen level.

1 = not at all
2 = rarely
3 = occasionally
4 = often
5 = always

	1	2	3	4	5
1 How often do you find that you stay online longer than you intended?					
2 How often do you neglect household chores to spend more time online?					
3 How often do you prefer the excitement of the Internet to intimacy with your partner?					
4 How often do you form new relationships with fellow online users?					
5 How often do others in your life complain to you about the amount of time you spend online?					
6 How often do your grades or schoolwork suffer because of the amount of time you spend online?					

	1	2	3	4	5
7 How often do you check email before something else that you need to do?					
8 How often does your job performance or productivity suffer because of the Internet?					
9 How often do you become defensive or secretive when anyone asks you what you do online?					
10 How often do you block out disturbing thoughts about your life with soothing thoughts of the Internet?					
11 How often do you find yourself anticipating when you will go online again?					
12 How often do you fear that life without the Internet would be boring, empty, and joyless?					
13 How often do you snap, yell, or act annoyed if someone bothers you while you are online?					
14 How often do you lose sleep due to late-night log-ins?					
15 How often do you feel preoccupied with the Internet when offline, or fantasise about being online?					
16 How often do you find yourself saying 'just a few more minutes' when online?					

	1	2	3	4	5
17 How often do you try to cut down the amount of time you spend online and fail?					
18 How often do you try to hide how long you've been online?					
19 How often do you choose to spend more time online over going out with others?					
20 How often do you feel depressed, moody, or nervous when you are offline, which goes away once you are back online?					

'The higher your score, the greater your level of addiction and the problems your Internet usage causes,' says Young.

20–39 points: You are an average online user. You may surf the web a bit too long at times, but you have control over your usage.
40–69 points: You are experiencing frequent problems because of the Internet. You should consider their full impact on your life.
70–100 points: Your Internet usage is causing significant problems in your life. You need to address them now.

NINE STEPS TO FOCUS

From now on, at the end of each chapter, you'll get a chance to check on your progress. There are nine steps to climb.

Well done, you're off to a great start by assessing your level of digital distraction. You've just climbed up to step one.

As you climb a new step at the end of each chapter, it's a good time to ask yourself a few questions about how things are going so far. This will help you notice what is working for you, and what isn't. It will also give you some ideas of what you need to work on as you move on to the next chapter.

STEP ONE

1. *As you take your first steps, what have you done to tackle digital distraction (for example, unsubscribed from an email alert, tailored your Facebook feed)?*
2. *Has anything changed as a result?*
3. *At this early stage, what are you finding tricky? What are you finding easier than you thought it would be?*

A QUICK RECAP

- It's time to start capturing information so you can build up an accurate picture of your digital distraction habits.
- Are you dangerously distracted, on a slippery slope, or mostly focused?
- In your digital distraction diary, note down the time,

what you were distracted by, and what you were distracted from.

- Keep your distraction diary for a week and then look for any patterns.
- Don't worry if you feel like giving up – that's normal at this stage.
- A lot of digital behaviour is tied up with feelings. Changing your thoughts from a negative to a positive cycle can make a huge difference to how you feel.
- Definitions for Internet addiction have been established. Even among those not addicted, a psychological impact of Internet use can be noticed.
- Start a habit of checking in at the end of each chapter to note down any attempts, discoveries and difficulties.

FURTHER READING

- Kimberly Young has more advice on how to tackle addiction in her book *Caught in the Net*: Young, K. (1998) *Caught in the Net: How to Recognise the Signs of Internet Addiction – and a Winning Strategy for Recovery*. New York: Wiley.
- If you feel you need help from a psychologist to cope with severe digital distraction or addiction issues, the directory of chartered psychologists on the British Psychological Society website is a good place to start: http://www.bps.org.uk/bpslegacy/dcp.

CHAPTER 5

Change starts today

A poor life this is if, full of care,
We have no time to stand and stare.
WILLIAM HENRY DAVIES

IF YOU ARE LIKE MOST people, chances are that, by now, you will have become addicted to, or formed automatic patterns of behaviour with at least one form of digital distraction.

You might have been surprised in the last chapter to realise that digital devices were playing such a big part in your life. This issue has crept up on many of us so don't worry if that is the case.

There's plenty you can do to regain control.

Smartphones, laptops and mobile devices have become part of our landscape – almost part of our being, in many cases.

We're not going to erase this whole aspect of the landscape – it would be totally unrealistic to think we would switch it all off and never use it again.

But, like when we move to a new city, we need to learn how to live here. We haven't really taken much time over this yet.

WHY FOCUS?

Once you taste focus, you're likely to want more of it.

So what does it feel like?

In a recent talk given by Joe Kraus, he talked about what it feels like at the point you are at your most creative:[1]

> Maybe it was your best round of golf, maybe it was solving a tricky computer science problem. Whatever it was, likely, you were LOST IN THE MOMENT, completely absorbed in what you were doing. It was long-form, not quick twitch. You were in the zone. Your attention was fixed, calm, present.
>
> Once people experience the zone, most of us want to get back there. It's a feeling of peak performance, peak creativity, peak aliveness.

Peak attention is another idea you are likely to hear much more about in the next few years. It's the opposite of scattered-brain thinking.

Suddenly you're getting things done. You've had an original thought because you are not letting anyone distract you. You're creating freely. You're producing volumes of work that were previously unthinkable.

You're calm, contented and stress-free. You're in an almost Zen-like state. Time passes unnoticed – a vague measure you are unaware of, rather than something you are always racing against.

This may well sound like nirvana compared with a buzzing, blinking, beeping, connected world.

Where is the smartphone here? Switched off. What about the Internet? Not connected. Emails? Unchecked so far today. And social media? You're going there this afternoon.

Focus is a skill we already had, but it has just become a whole lot more inviting. It sits in stark contrast to the state of being constantly connected and distracted. It offers us a free, instant ticket to boost our productivity.

Ready to grab it?

WHAT IS FOCUS?

Have you ever looked up at the clock and realised that the hands have spun round without you even noticing because you were so absorbed in what you were doing?

What were you doing at that point? Chances are, it wasn't checking your email...

When time passes without us even noticing, we're in a state called flow. This is a calm, productive place to be. We're unlikely to be distracted and, indeed, tend not even to notice what is going on around us. We're utterly focused on the task.

Flow is good for your health and stress levels, and excellent for productivity.

This is being focused.

But digital distraction and focus don't just amble along happily together. It's one or the other. You can't be focused and digitally distracted at once.

In fact, distraction can be seen as the opposite of focus, as Leo Babauta explains:[2]

> A new email comes in, and so we must stop what we're doing to check the new email, and possibly respond. Even if we don't respond right away, whatever we were just doing was interrupted.
>
> This is the opposite of focus and nothing exemplifies the need for focus better. Sure, you're always in touch, always up to date, always on top of things. But you have

no focus, and you're buffeted in all directions by the winds of your email (or Twitter, Facebook, IM or other communication channels).

Distractions knock us off course, and deplete our productivity. One of the biggest problems is that we never know when they're coming. We do have control over this but we don't make use of it – we need to learn how to use the 'off' button more readily.

FIND YOUR POWERFUL POINT OF FOCUS

What – for you – creates a feeling of being in flow?

Is it gardening? Painting? Reading? Do you lose yourself in DIY? Photography? Running?

Think about what it is that, for you, makes time pass without you noticing. When are you most absorbed?

Write down a few ideas that spring to mind:

1
2
3

Can you think of any point in the last week when you felt focused?

Next time you notice yourself being focused, think about how it makes you feel.

- *Notice your breathing pattern, which will typically be very regular and relaxed.*

- *Notice whether you are feeling stressed or not.*
- *Notice how content you feel, on a scale of one to ten.*

Start today

> *Try and spend some time today doing one of the things that create a sense of flow for you.*

Let's look at the opposite as well – what does distraction make you feel like?

Next time you catch yourself being digitally distracted, stop for a moment.

Think about how you are feeling. Ask yourself the questions below, a variation of the exercise for flow.

- *Notice your breathing pattern.*
- *Notice whether you are feeling stressed or not.*
- *Notice how content you feel, on a scale of one to ten.*

Do you feel very different when you are distracted from when you are in a state of flow and focus?

Evidence shows that people tend to be happiest when absorbed in a task, in flow. People are happier pursuing a goal that they have to concentrate full attention on (when they are lost in the moment) than they are at many other times when you might expect that they would be happier, such as when relaxing.[3]

What the mind is doing is very different in these two states. Digital distraction is the mind flitting around, often overloaded. Flow is the mind absorbed on just one thing.

Roland Jouvent is a professor of psychiatry at the Paris VI University and director of the Centre for Emotions at France's Centre for Scientific Research.

He says a racing mind causes exhaustion. He suggests slowing down our minds by doing simple things – when walking, focusing on every step, for example. He talks about the impact of technology on the mind:[4]

> The more the mind accelerates, the more it is disassociated from the body, and then it becomes exhausted and lost. We struggle to slow down because the speed of new technology – its images, words and ideas – removes us from the present. It's addictive.

Swap shop

The trick now is to replace distraction with focus bit by bit.

You need to take the focus you have experienced (be that while running, painting or singing, for example) and transfer it across to day-to-day life.

In day-to-day life, we are often not as easily absorbed as we are during our favourite pastimes.

When you try and focus, these are some things that are likely to happen:

- You have an urge to check email.
- Your hand moves automatically to click an icon to open social media/email/Internet.
- You feel like the task you are meant to focus on is too difficult.
- You think about not carrying on with the task.

When we are bored, or find something difficult, we are more likely to be distracted. But we also know that distraction breeds distraction and, so, if we can avoid distractions, we will be able to settle in to focusing on our task.

Try again, this time while switching off one digital distraction for 15 minutes (so switch off either your smartphone, email, social media, or close down your Internet browser for 15 minutes).

Why do you need to do this?

It's extremely difficult to resist the pull of say, an incoming email, when it is right there in front of you. So switch it off, just for 10 or 15 minutes and focus on your task.

Then focus on what you are doing.

Notice how it feels.

Keep going. If you get distracted, don't worry about it, just note down how long you stayed focused for.

The next day, add in switching off another digital distraction, too – for example switch both your email and your social media off for 15 minutes.

By the end of the week, decide whether you could switch off all digital distractions (including your phone) – even for just 15 minutes. If you think you can, then do so.

Next time, set yourself the target of staying focused for slightly longer. Build it up gradually – from 15 minutes to half an hour.

Notice what you are getting done. Notice how you feel when you are focused. Notice your breathing and how content you feel. Does time pass without you noticing?

Identifying what focus feels like and knowing you can focus even for a short time is a great start.

Don't try and switch everything off for hours at a time straight away if you are highly dependent on digital platforms. You'll be likely to have withdrawal symptoms if you do.

FIND YOUR TIME OF FOCUS

Now you've identified what focus feels like for you, the next step is to give yourself the best possible chance of achieving that focus each day.

Our bodies go through natural rhythms of energy highs and lows every 24 hours. These are different for each person.

If you learn to recognise your natural energy patterns, you can start to make the most of your productive times.

Say you realised that your most productive time was 9 am each day. But at that time, usually, you spent an hour checking and replying to your emails. What a waste!

Avoid digital distractions at your most productive time, if at all possible. That's a precious slice of the day when you could be devoting great energy to making quick progress on focused tasks.

When do you have the most energy?

Think about which point in the day you feel full of energy, and which time of day you feel like you can't be bothered with anything, in a slump (that feeling when you're yawning and don't want to do anything...).

Some people's bursts of energy will last one or two hours, once, twice, or three times a day. Everyone is different, as shown in a presentation by digital wellbeing expert Sinead Mac Manus (see Chapter 11 for more tips on focus from Sinead).[5]

Keep a check today and tomorrow of your energy highs and lows.

What should you be doing during these energy highs and energy lows?

Prime-time activities

- Creating
- Writing
- The most difficult task of the day
- The most important task of the day
- Planning
- Idea generation
- Getting inspiration (for example, by going for a walk)
- Important meetings
- Difficult tasks
- Reporting
- Learning.

Slump-time activities

- Checking email
- Digital distractions
- Shutting your eyes for 10 minutes
- Having a chat with a colleague
- Making a cup of tea

- Less important meetings
- Chores
- Easy tasks
- Taking a break
- Restoring energy.

Something has to fall into our slump times of day – we can't be highly focused all the time. But if you're conscious of your body's energy highs and dips, you can give yourself a break when you hit a slump.

There's plenty of time for digital distractions in your least productive points of the day (they're likely to be more of a temptation then anyway). A good time to do a bit of emailing, check Twitter, or browse the Internet is when you are in a slump. You can do so safe in the knowledge that you are not wasting your most powerful time of day.

If you don't have to work in your least productive times, then don't. Save your energy for when it counts.

If you have to sit in an office for eight hours a day, you could argue that you have little say in what you do. It may seem that you must just work through the low-energy times. But there are still ways to use this knowledge to your advantage.

Organise your time so you can be doing your most difficult task at your most productive time – and during this time, don't let anyone interrupt. If they try to, ask if you can get back to them tomorrow or this afternoon, then make a note of it, and carry on.

Meetings are a tricky area. They sometimes spread slump. If you suspect the meeting will not be productive – and you can't get out of it – don't schedule it for your most productive time. On the other hand, if you're likely to come out of the meeting buzzing with ideas and inspired, book it for a great time of day for you.

Some activities can restore us at any time of the day. Walking is a classic example, as is meditation. We'll look at how to restore and refresh in depth in Chapter 11.

IDENTIFY WEAKNESSES

Everyone has weaknesses. It is useful to pinpoint yours, so you know what you are dealing with. Tick your answers.

What is your worst digital distraction?

	✓
Email	
Social media	
Internet	
Smartphone/mobile	
Something else (name, if so…)	

What time of day are you most likely to get digitally distracted?

	✓
First thing in the morning	
Mid-morning	
Lunchtime	
Afternoon	

Evening	
During slump times of day	
Any time	

Do digital distractions interrupt you most:

	✓
When at work?	
When travelling?	
When with others?	
When alone?	
When you first switch on your computer?	
When at home?	

What kind of mood/state are you usually in when you get digitally distracted?

	✓
Bored	
Tired	
Reluctant to do a difficult task	
Procrastinating	
Angry	
Happy	
Lonely	
Scared (of whatever task is ahead)	

This is useful ammunition. You can be extra watchful of your behaviour when using certain platforms, at certain times of day, or when you are in a certain mood.

If you are not sure about your answers to any of these questions, look back at the digital distraction diary you kept in the last chapter. Look at the patterns you noticed, and use these to help you with your answers.

What is life like as a #addict? Twitter super-user, Angela Clarke, tells all…

I tweet 28 times a day, on average, on my personal Twitter account (so a phone app tells me). I'm a writer, and I also have a work account, which I set up to promote my latest book. Here, I tweet 11 times a day. Eleven tweets for work seem reasonable. That's only slightly more than Twitter deity, Stephen Fry's, eight and a half tweets.

There's no repetition between the two accounts, so I post 39 times a day in total. The average tweet contains 15 words, which means I write… 585 words per day on Twitter. That's a quarter of a book chapter. That's a whole newspaper column. I have a problem.

The first thing I do in the morning and the last thing I do at night is check Twitter. Twitter is the reassuring background chatter to my day. I'm freelance and I work from home. Like an office worker may turn to their colleague at the next desk and share their thoughts, so I turn to Twitter. I post about my frustrations, my happiness, and my observations on the pitfalls of eating a sugary doughnut over your keyboard (v. crunchy). And I do it all while wearing pyjamas.

I love my imaginary friends (followers sounds too much like a cult). Our friendship is not based on geographic convenience, or built around going to the same school/university/workplace. We're mates because we enjoy the same interests, whether that's discussing the merits of the Booker Prize Shortlist or commenting on Mary Berry's buns on Great British Bake Off.

*My imaginary friends alert me to the latest news, cultural phenomenon, or humorous YouTube video of a kitten falling in a toilet. Twitter doesn't comment on the Zeitgeist, it *is* the Zeitgeist. Because of Twitter I stopped everything to watch the Chilean Miners be rescued in 2010, the final US Space Shuttle Atlantis launch, the England Riots in 2011, and pretty much every second of the London 2012 Olympics. I'll be able to tell my grandkids I was there, laptop in hand.*

The problem is, while I was posting 140 character comments on historical events, I wasn't working. Think of all the things my imaginary friends and I could have done if we weren't mucking about on Twitter! Instead of commenting on history, we could have made it.

Writer and author Angela Clarke tweets
@TheAngelaClarke

Exercise to begin total focus on a task today

---

1 Think of the thing you want to focus on doing – write it down.
2 Set aside a time when you are going to start on this – put it in your diary.
3 Tell anyone you urgently need to tell that you will be unavailable by email/phone during that time.
4 Switch off all digital distractions (email/mobile/Internet/social media).

5 Begin task.
6 Do the task for an hour if you can.
7 See how much you get done.
8 Schedule in the next time you can focus again on the task.
9 Switch back on and re-join the digital world.

Each day ask yourself how long you would like to spend on focused tasks.

- *Are there any barriers to focusing today? Are there people or things that might get in the way? If so, list them.*
- *What is your aim today?*
- *Are there digital tasks that you need to do? If so, what are they?*

We'll return to look at focus again in Chapter 8, where you'll build on the work you've started here.

But first, a bit of inspiration…

In the next two chapters, you'll get tips from productivity thinkers, and see how the pros battle digital distraction.

NINE STEPS TO FOCUS

You've made it to step two, well done.

This is a good point to think about what's working well, and what is proving trickier. Let's check on your progress.

STEP TWO

1. *Are you feeling focused, or are distractions still tempting you?*
2. *What new ways have you tried to tackle digital distraction? Did they work?*
3. *What happens for you when you are focused?*

A QUICK RECAP

- Think about a time when you were really absorbed. Being in a Zen-like state where time passes and you can create freely is worth aiming for.
- Focus and flow are productive states. Distraction is the opposite of focus.
- Identify what focus feels like for you, through an activity such as painting or gardening. Notice how different this feels from distraction.
- What is your best time of day? Think about slotting important activities into prime time and less demanding activities into slump time.
- What are your weaknesses? Know these and you'll understand your habits better.
- Focus on a task today. Notice what a difference lack of distraction makes.

FURTHER READING

- This lecture by Joe Kraus is great for a bit of inspiration if you've only got 15 minutes. You can watch it or read the transcript. Kraus, J. 'We're creating a culture of distraction', http://joekraus.com/were-creating-a-culture-of-distraction, 25 May 2012.
- Leo Babauta is one productivity thinker I'd recommend getting to know through his writing. His books and blogs are a pleasure to read and full of sound advice. This free ebook is a great place to start: Babauta, L. (2010) *Focus: A Simplicity Manifesto in the Age of Distraction*, ebook.

CHAPTER 6

Watch the pros

Morning air! If men will not drink of this at the fountain-head of the day, why, then, we must even bottle up some and sell it in the shops, for the benefit of those who have lost their subscription ticket to morning time in this world. HENRY DAVID THOREAU

IN THIS CHAPTER WE'LL LOOK at the lessons we can learn about focusing from the world of business, the world of sport, and influential thinkers past and present.

The distractions we battle against today – email, the Internet, social media and smartphones – are a particularly modern curse.

But distraction itself is nothing new.[1]

A BlackBerry or an iPhone would certainly look quite out of place in the 1800s, the time of Henry David Thoreau.

But that doesn't mean that ideas thought up by Thoreau then cannot be borrowed today – in fact some of them are more useful in our constantly connected world than they were then.

Thoreau retreated to the woods near Walden Pond for two years, away from all company, and connected with his thoughts and the environment.[2]

What Thoreau would have made of our need to update each other every two minutes with trivial minutiae via social media or messaging, I dread to think.

He felt that, even back then, people were in danger of not having enough of value to say to one another: 'Society is commonly too cheap. We meet at very short intervals, not having had time to acquire any new value for each other. We meet at meals three times a day, and give each other a new taste of that old musty cheese that we are...'

The lesson from Thoreau? If we step away in a connected world and experience something for ourselves, then there will be more to share.

MEET THE PROFESSORS

That's all very well, you may be thinking, but in today's world? As if!

Well if you don't fancy camping beside a pond, how about retreating to a remote area of southern Utah, USA instead?

This is exactly where a modern-day Thoreau – professor David Strayer – led an expedition of neuroscientists in 2010.

His aim? To find out how heavy use of digital devices and technology changes how we think and behave, and to see how the brain is changed by a retreat into nature.

Strayer, a professor of psychology at the University of Utah, sees attention as 'the holy grail'.[3] 'Everything that you're conscious of, everything you let in, everything you remember and you forget, depends on it,' he says.

He believes that too much digital stimulation day-to-day can 'take people who would be functioning OK and put them in a range where they're not psychologically healthy'.

To see just how beneficial it was to switch off entirely for three days, he took with him four other brain experts – some of whom were sceptical. They all switched off entirely from mobiles and the Internet, as they trekked and rafted down the San Juan River.

The scientists experienced a 'mental freedom' from knowing they could not be interrupted by anyone and anything and, as the days passed, they did plenty of important thinking.

Because they were all together, and experts in their field, they were able to discuss their thinking, pushing their ideas even further. Topics they thought about included how to study the toll taken by constant interruption from email, and whether merely anticipating email knocks attention and focus.

They took other questions home to research, including: Why are people distracted by irrelevant streams of information? Why do teenagers decide to text in dangerous situations? How does the brain reset as if in nature?

These brain experts know better than most what an impact technology has on the brain. The example they set – taking a total focus break in nature and leaving all digital distractions behind, is one we can follow.

They have demonstrated to us that the brain needs to be reset.

INTO THE BOARDROOM

That's all very well for brain scientists, you might say, it's practically their job to switch off.

It could never happen in the cut-throat corporate world.

Could never happen? In that rat race where people sleep with their smartphones, and emails are your stock in trade?

One researcher has proved otherwise.

Professor Leslie Perlow went in to a major consultancy company and found that once people disconnected for a few predetermined hours every week, they worked more productively, and were also happier about work.[4]

The company Perlow chose was the Boston Consulting Group (BCG). She deliberately chose a hard environment, figuring that, if change could happen there, it could happen anywhere.

Many people actually thrived on the intensity of the work there, and didn't want it to change, she found. They accepted the demands on their time as the price they had to pay for high salaries.

In her book, *Sleeping With Your Smartphone*, she gives an account of what a difference switching off made.

Initially, a survey of 1,600 managers and professionals showed that one quarter of them slept with their phone. Some 70 per cent checked their phone each day within an hour of getting up, and more than half checked continuously when on holiday. Most reported spending 25 hours per week monitoring work on top of what was already at least a 50-hour week (for most).

'Simply put, people were "on" a great deal,' says Perlow.

Consultants found that, by having predictable time off (PTO), and by helping each other take this time off, their lives and their work results improved. Within four years, more than 900 teams from 30 countries had taken part.

Perlow's results show that those taking part were more likely to be excited about starting work in the morning, were more satisfied with their job and their work–life balance, and were more likely to rate their team as collaborative and effective.

She uncovered a valuable lesson about why things were going wrong before. People were responding to demands – and being highly available – and so these demands increased.

Perlow calls it a 'cycle of responsiveness'. At the beginning, she says the pressure to be 'on' stems from 'some seemingly legitimate reason', such as people in different time zones.

People adjust to these demands – altering their schedule, adapting the technology they use, changing the way they live with friends and families – to meet the increased demands. But once clients get increased responsiveness, they ask for more. Most workers – already working long hours – accept these additional demands (urgent or not).

> Eventually, the cycle grows (unintentionally) vicious; most people don't notice that they are spinning their way into a 24/7 workweek. And even if they begin resenting how much their work is spilling into their personal lives, they fail to recognise that they are their own worst enemy, the source of much of the pressure that they attribute to the nature of their business. LESLIE PERLOW

When we're there, people ask things of us. Often they do this rather than thinking for themselves, or trying to answer their own questions. I often find that if I don't check email for a while (a few days) then I'll have two emails from the same person – the first asking me to answer something, the second saying they've solved it themselves and not to worry.

It's the equivalent of (in person) people saying 'Ah, just the person' or 'I was meaning to ask you'. If people can see we are available, they will ask.

Think about it – can you remember a point where you adjusted your behaviour to accommodate an initial demand (such as checking email at an antisocial hour)?

What happened next? Did you snap back to your original email policy or did the elastic not ping back fully?

If you check email or social media first thing in the morning, can you remember how or why you started doing that?

Perlow points out that working in a way where you are not so accommodating and there is not a pressure to be always 'on' is more efficient, more effective, and better for the individual and the employer.

What can we take from this? Don't assume you should always be 'on' – even if the pressure to be so feels immense. Plan predictable time off.

REPLY TO ALL: TAKE A BREAK

Another office where the tables have been overturned when it comes to digital policy, is that of information technology giant, Atos.

The chief executive, Thierry Breton, has announced that internal email is to be banned among his staff from 2014.[5]

So no more chitter-chatter with Danielle four desks along and jokes with Tom from admin forming the brush-strokes painting your day.

What will they do instead?

Work, says Breton.

'I started to think they were spending too much time on internal emails and not enough time on management,' he said.

Atos employs 80,000 staff in 42 countries. Breton found out that, on average, each of them was receiving more than 100 internal emails per day.

With more analysis, Atos found that employees thought only 15 per cent of their messages were useful – meaning processing the rest was lost time.

Yet the staff were spending 15 to 20 hours a week checking and answering emails.

When the numbers are laid out for inspection like this, you wonder why every company on the planet isn't adopting a policy like Breton's.

Breton talks of a 'data deluge', saying the data a company collects and stores doubles every 18 months, and that 'more data was created over the past 10 years than since the beginning of humanity'.

Internally, Atos will switch to tools including instant messaging, social networks, and cloud computing. They will keep external emailing.

This stance is a brave one. It is worth noting it has been taken by an information technology company – at the cutting edge of what they are dealing with.

They are likely to be front runners in finding a solution to some of the technological ills companies are starting to experience – such as 85 per cent of internal email being pointless.

We can use their ideas to challenge ourselves, and to think before we hit 'send': do we really need to send this email?

Another example of those who work closely with the web being aware of its distractions is web design company, Top Left Design. Their designers go into a 'design bubble' to focus. Keren Lerner explains.

Our work requires us to concentrate on things like design and coding for a few hours at a time. But we also work with clients, and so need time to respond to emails and manage team projects too.

How do we combine the two?

Each designer is responsible for between five and 11 live projects. They project-manage, and communicate directly with clients, so we can get a better understanding of what clients want.

We need to keep on top of our emails and speak on the phone to clients, and we need to communicate with each other and our suppliers about projects.

However, our designers also need concentrated time to be able to focus on the research, image sourcing, layout and design of the pages for each website. It's on these occasions where we need the 'design bubble'.

How the design bubble works:

1 *One of the team needs to work on design work uninterrupted for two to three hours. So, he or she tells everyone else, 'I am going into the design bubble!'.*
2 *Then they put on their Out of Office in their email, and put their headphones on.*
3 *The Out of Office tells clients that they are otherwise engaged, and if the phone rings for them, one of the other team will take the call and take a message for them. So this means they can work uninterrupted and create their amazing designs.*

The same applies for our coders – they need to concentrate, too, when they code and test our websites.

When we moved to new offices, we even made one of the rooms into a design bubble.

Team members can physically go and work in that room, so they really are alone. But most of the time, the headphones/Out of Office works just fine and people stay at their desks.

That's the story of the design bubble. It's a system that anyone can use – and if you do we would love to hear about it. Send us a tweet @topleftdesign or write on our Facebook wall: www.facebook.com/topleftdesign.

TAKING GOLD

From the boardroom to the Olympics – and a piece of no-nonsense advice from boxer Amir Khan.[6]

> Turn your phones off. Your job is to win that gold medal. AMIR KHAN

Right. It's off. Gold medal here we come…

This advice was given by the professional boxer during the London 2012 Olympic Games to athletes going for gold.

An awareness of the need to avoid distractions was also shown by some Great British athletes who posted on their website that they would be 'trying to avoid distractions such as social media' while they focused on delivering their best performance at the Olympics.[7]

Focus is something that has long been associated with the sports field – perhaps because it is one arena where it is so visible.

Think of the 100 metres final. One sprinter renowned for having tunnel vision was Linford Christie. His mind was almost visibly fixed on his goal as he went for (and won) gold in the Barcelona 1992 Olympics.

'Barcelona was the first time I went to the blocks imagining my lane was a tunnel, with everything else on either side a blur,' he wrote.[8]

Often, in the sports arena, psychologists help athletes develop mental excellence, as part of a team that includes coaches, trainers and nutritional experts.

The mental side of the equation is not ignored on the sports field – something those of us who are not athletes can learn from.

One sport psychologist who has coached Olympians is Simon Hartley. He points out how important focus is and says that we can only focus on one thing at a time. He says he often describes focus as a torch beam:[9]

> The challenge for many athletes is actually to take control of the torch and decide where the beam will shine. What is it that you need to focus on?... A lot of us don't have the mental discipline required to hold our focus for any prolonged period. We are easily distracted and find that something draws the torch beam.

Even if we have no intention of putting in enough physical hours of training to make the Olympic stage, mental toughness is something we can learn.

ALL AROUND YOU

Look around you, and you'll start to spot examples of focus in some unlikely places, as well as places you'd expect.

We expect pilots, surgeons and air traffic controllers to focus. But can you spot other people demonstrating focus too?

As you spot examples of focus in expected, as well as unlikely places, note your observations down in your notebook.

NINE STEPS TO FOCUS

Well done! Step three! How is your battle against digital distraction going? Perhaps, by now, you've started to notice a change in how you act.

STEP THREE

Ask yourself these questions:

1. *Did the examples in this chapter give you any ideas that you might try?*
2. *What is working well for you in tackling digital distraction?*
3. *What do you feel you still need to work on?*

A QUICK RECAP

- Thoreau stepped away from other people to have a long and proper think. What are the chances of doing the same these days?
- A group of neuroscientists who switched off while discussing the brain in a remote area of Utah can remind us of the importance of resetting the brain (especially while in nature).

- Attitudes changed in the Boston Consulting Group when consultants realised the benefits to themselves and their clients of scheduled time switched off.
- Do you use internal email for pointless emailing? What would happen if your boss switched it off?
- The Olympic 100 metres final is one event where focus is clearly visible. Like athletes, we can prepare mentally to focus.
- Can you create your own 'no-digital' bubble? What could you create while you are there?

FURTHER READING

- For the full story from Leslie Perlow, see her book (2012) *Sleeping With Your Smartphone: How to Break the 24/7 Habit and Change the Way You Work*. Harvard Business Review Press.
- Read more about what Henry David Thoreau learnt during his experiment in living in his classic (1962) book, *Walden, or Life in the Woods: On the Duty of Civil Disobedience*. New York: Macmillan.

CHAPTER 7

Tools and methods (productivity tips)

This is the Information Age, which does not always mean information in our brains. We sometimes feel that it means information whistling by our ears at light speed, too fast to be absorbed. JAMES GLEICK

THE BIGGEST TOOL AT YOUR disposal in fighting digital distraction is willpower.

But, for even the most focused among us, there are points when this wanes.

When that happens, it's useful to be able to trick yourself into staying focused.

Different things work for different people. And what works one day might not work another day in a different situation, so keep a couple of ideas as a backup.

These methods include using technology, relaxation techniques and goal setting. Hopefully there is something among them that appeals to you.

These are methods to help you with your own motivation and thinking, for the times you are playing your own worst enemy and it's your own behaviour that's making you turn to digital distraction. Many of the tips are good for productivity in general, as well as battling digital distraction.

We'll look in the next two chapters at what to do when it's other people causing the distraction.

In need of a shot of willpower, or a new slant on how to get focused? Here are 15 quick tips.

1 DEADLINE DRIVEN

There's nothing quite like a deadline to force focus. I watched this method motivate hundreds of employees, day after day when I worked in newspapers. The paper has to go to print, the presses will roll, and a big gap at the top of page 6 (or any other page!) is really not an option.

While racing to deadline constantly is not particularly healthy as a way of life, we can certainly borrow this method now and again.

Say you had just 20 minutes to go before the bell rang for the end of the day? What would you do in those 20 minutes? Probably not check Facebook… Try telling yourself there are just 20 minutes until deadline when you feel your focus slipping.

If you don't set many deadlines, experiment with setting yourself a few more. Then tell people you're on deadline when they come to you with unimportant requests – it's true.

If you can't convince yourself that a deadline you set for yourself is real, then get someone (for example, your boss) to set one for you.

You might well have heard of Parkinson's Law – that work expands to fill the time available.[1] That's why a deadline – that has to be met – helps.

2 EYE ON THE PRIZE

Why shouldn't you win prizes every day when you're battling against digital distraction?

Reward-setting is a great way to keep yourself going on a micro scale – for the next half an hour, then the hour after that.

The rewards don't have to be anything fancy or expensive – they can be as simple as 'I'll put the kettle on in half an hour once I've done that' or 'I'll have a walk and some fresh air for 20 minutes at lunchtime'.

These rewards can also be made up of exactly what it is you are trying to avoid. So you can decide to have a 'Facebook lunchbreak' or agree with yourself you'll spend 30 minutes at the end of the day on Twitter.

3 MOST IMPORTANT TASK

This is a method created by Leo Babauta, who advises setting yourself up to three Most Important Tasks to be the focus of each day.[2]

You should set these each morning and get on with them straight away.

Doing them first is really important – you need to use the energy you have at the start of the day for them. 'If you put them off till later, you will get busy and run out of time to do them,' says Babauta.

If you start the day by checking your emails instead, you get sucked in to responding to other people's demands (their important tasks, not yours). You could get to lunchtime having done lots of small unimportant tasks, but nothing truly important.

If you feel like you're overwhelmed and busy, but never actually making solid progress on important things, then set Most Important Tasks each day and do them. Don't let small unimportant tasks take priority.

4 DELEGATE YOUR DISTRACTION

Had enough of being digitally distracted all the time? Then why not delegate it?

In his book, *Future Minds*, one of the predictions that author Richard Watson makes is that, in the future, we will employ information sifters to do the dirty work of digital distraction.[3]

An information sifter could check your email and weed out everything you didn't need to know, do your Internet searches, and scan your social media accounts. It would be their attention being used up on these things, not yours.

> *If you could delegate all your digital platforms, would you take the option?*

As with any form of delegation, it's a big step to relinquish control. But, in the very near future, the role of PAs and VAs (virtual assistants) will most likely expand to include much more information sifting.

5 TAKE OUT A TASK

There are so many things on your To Do list, they're all squeezed together, elbows tucked in...

You can feel the panic rising, and your stomach

tightening. You're running here, sprinting there, but getting nothing done.

When there's too much on your list, it's easy to get overwhelmed and achieve nothing. You spend your energy thinking about how you can't possibly get everything done, rather than doing any of it.

Get ruthless. What are you going to elbow off the edge of the list? Sometimes this means tough decisions. Prioritise.

A word of warning, however, when you're choosing what to elbow out of the way – don't just choose the thing you're put off by.

It may be that many of the other tasks have materialised only so you can put off this one thing a bit longer. 'Eat the frog', as an old saying goes – get the worst, most challenging thing done in the morning, the task you are most likely to procrastinate over. Everything is easy after that.[4]

6 RESULTS ONLY

A huge multiplier of digital distraction is presenteeism. When people are at their desks only to clock up the hours and show their faces, they are likely to get distracted once they get bored or hit an energy low.

A different approach is to focus instead on results. Results Only Work Environment (ROWE) is the philosophy that what you get done at work is more important than the time you spend there.[5]

Sitting at your desk from 9 am until 5 pm is an unproductive way to work – during some of this time your energy and productivity will be very low, and you are likely to get digitally distracted. Instead, you could rest, restore your energy, then get things done even more quickly at productive times.

Many offices have a culture of presenteeism. If yours is one of them, you may well find it de-motivating. Talk to your manager about focusing on results and setting goals, rather than the emphasis being on just showing up. You'll be doing them a favour. By having a ROWE approach, it means when people are present they are actually being productive.

7 CLOSE THE LOOPS

In his highly influential productivity book, *Getting Things Done*, David Allen talks about 'open loops' which sap productivity.[6] These are things pulling on your attention – anything you are aware of that needs to be done.

By closing these loops we increase our productivity because we no longer have to keep a running mental note for ourselves, 'I must do X'. It's about finishing, so you can clear through one task after the next.

Having seven different Internet browser windows open at once, and attempting to multitask across three projects is not closing loops.

Get one thing done – close the loop – then start the next. One problem Allen points out is that these days 'for many of us, there are no edges to most of our projects'. Before, in a factory for example, you could see when a box was packed and finished.

But when is a project finished these days – especially when we carry it home with us on our smartphones? We need to draw more lines to give ourselves more headspace.

8 STOP THE CLOCK

It used to take days to write a letter and send a reply. Then came email. We got faster, and started dealing in hours, then minutes. Then we adopted instant messaging, text messaging and micro-messaging. We started dealing in seconds, then in instant.

Next, perhaps we'll respond before we've received the message. Dealing in empty thoughts is not as ridiculous as it sounds – after all, many of the thoughts we send now are vacuous, because we leave ourselves no time to think them through before hitting 'reply'.

In his book, *Faster*, James Gleick says that, when speed of correspondence increases, early adopters have a competitive advantage. Soon, though, everyone adopts the new timescale and equality is restored. The only change that remains is a universally faster pace.[7]

In the past, as Gleick notes, '... Business ran like correspondence chess, with plenty of time for contemplation'. 'We may need to set aside formal time for deliberation, where we once used accidental time,' he advises.

We can't go any faster now. The competitive advantage comes from going more slowly. Build in time for contemplation, and get the edge.

9 CHECK YOUR FILTERS

> We are to information overload as fish are to water. It's just what we swim in. CLAY SHIRKY

As Shirky points out, information overload is nothing new – we have had it as part of our lives since the printing press was invented.[8]

When we feel we are suffering from information overload, what has gone wrong?

Shirky says ask yourself which filter just broke. What were you relying on?

He points out that the filters we did have in place – to regulate things such as privacy and information flow – are breaking as they are challenged by new demands. We are re-thinking social norms, in how we interact and communicate. But we are relying on old filters.

One of the most straightforward applications of filters is for email. Here, a holiday is often a good way to reassess. When you get back, you will have a clearer perspective of which messages to filter out.

If you are brave enough, you could even pass your email account over to a boss or colleague, and ask them to set up filters for you while you are away, on emails you are getting that they think you don't need to see.

10 LOWER YOUR EXPECTATIONS

As Steve Parks, co-founder of web consultancy WunderRoot, notes in the following case study, we've got an expectation problem.

We now expect an instant response from each other. But, worse still, we expect an instant reaction from ourselves.

We also expect ourselves to operate across multiple platforms and always be up to date.

It's a lot to ask.

We're quick to blame others for our overflowing inboxes and time lost to social media, but sometimes we need to turn the mirror on ourselves. Are we the ones turning the pressure up?

Simply by lowering our expectations, we can cut out the seeming need to spread our attention so thinly. Often, the highest expectations are the ones we set ourselves. You post on social media every 15 minutes because you feel you should. What happens when you break that 'rule'? Nothing.

Before clocks existed, we based our rhythms on the earth and our bodies. We're so out of touch with this, it's frightening. The new rhythms we respond to are taxing our attention.

Think deeply and take time, don't race against it.

Steve Parks is one web-savvy professional who has chosen not to be constantly connected, so he can focus on the here and now. This is how he does it.

As a self-confessed geek, and co-founder of a web agency with 140 staff in nine countries, you might think I'd be addicted to being constantly connected.

But I'm a big believer of really being where I am, and with who I'm with. As a society we are allowing our gadgets to become electronic bungee ropes that ping us from where we are to somewhere else. I think that's disrespectful to others. It also leads to us becoming less productive, and less happy.

So, I've developed ways to use digital communication to fit around my priorities, rather than the other way around.

My mobile phone is nearly always on silent. I check it at times that suit me for messages. I schedule calls I know I want to have with people.

I have concentrated periods of work, where even my desk phone gets diverted to voicemail. No distractions equals better work. At the start of each day I also set the three most important things that I really must do, and no interruption is allowed to override those.

Email is a great communication medium because it's asynchronous. That means we don't need to check it every few minutes. But it has become an expectation, a demand, to be always available instantly, and that replies must be immediate. Just 10 years ago you'd send someone a letter and expect a reply around a week later. These days you send someone an email and expect a reply within the hour.

To deal with this I've switched off all notifications on my computer that show when new email arrives. I have all my communications programmes minimised, and only open them when I decide to have a comms session. During these sessions I either do quick replies, or I create a task in my electronic To Do list with a copy of the text of the email, and some notes on what I need to do – perhaps sending a holding reply to the person concerned, and archive the email itself to remove it from my inbox. I also set the level of expectation on response times, by normally not replying the same day.

As well as taking these steps myself, I expect certain things of others. I expect them not to check email or texts while we're chatting, and I don't expect them to reply to my digital communications immediately.

These steps help me to really be where I am. I think that's very important.

Steve Parks is a co-founder of WunderRoot, an open source web consultancy. He is also the author of a series of business books for Pearson. For more information see www.steveparks.co.uk.

11 BATCH TASK

A classic productivity tip is to batch task the things you have to do, grouping similar things together. Batch tasking works well for emails, phone calls and paperwork.[9]

You are less likely to fall prey to constant distractions if you do this. For example, group together all the emails you need to send, and do them all at once at 2 pm, rather than logging in to your email once at 9 am, once at 11 am, then again at 1.30 pm. While you are there in your inbox, other emails are likely to tempt you off-task and distract your attention. Better that this happens only once rather than three times.

Once you are already doing one sort of task, it is quicker to do another task of the same sort, without repeating the building up to it and setting up time (think of ironing, for example, or getting ready and getting out of the house).

The other productivity saving by batching similar things together is of time spent in switch-tasking (we'll look at switch-tasking in detail in the next chapter).

12 WATCH THE CLOCK

Where did our day go?!

Being unrealistic about how long we are going to spend on digital distractions is a classic trap most people fall in to. 'I'll just check…' turns into an hour.

Is it all just some game being played with our time? Things seem so fast online, we're whizzing from place to place, then we land back in our own front room and the hands of the clock have been spun round and round.

The Internet is the biggest time drain invented in our lifetime.

Why do we tend to so frequently underestimate how long something will take us on email/social media/the Internet? Because we don't add in the distraction factor.

The task we went there to do might take us exactly as long as we expected. But then we get distracted (by other emails in our inbox or by web links, for example) and use more time.

To begin to adjust your expectations to match reality, you need to keep an eye out like a hawk. Watch that clock, and note down the time that you log in to a platform. Set a limit on how long you want to spend there. When time's up, log out. You'll gradually start to judge better how long a task online will realistically take.

13 GOAL SETTING

Sometimes, we get so overwhelmed by a huge life goal (say, finding a new job) that, instead of tackling that monumental task, we go off and check Facebook 17 times in two days (which is far easier). Distractions tend to hit particularly when we are faced with a difficult task.

The way to beat this is to break things down into manageable chunks – setting goals step by step rather than one big end goal. Combine this with rewards (which can include digital media) as you reach each step of the goal.

So, goal one for the task above might be: list 10 sorts of roles I would like to work in.

Once that's done, set the next step on your route to your goal. Keep thinking 'what's next, what's next?' You should always know what is coming up on your goal steps list.

One of the trickiest times to avoid distraction when tackling these steps is when you first sit down to work on them (the start of each day is often a time people are tempted by distractions). The only way round this? Just start.

14 TECH SUPPORT

You don't have to conjure up all the willpower yourself. There are programs and apps that can help you battle digital distraction.

These are just a few of a wide selection:[10]

- Freedom: A tool for offline productivity, blocking Internet access on a Mac or PC for up to eight hours.
- Anti-social: Productivity app for Macs that blocks access to social media.
- SelfControl: A Mac app where you choose what distracts you (e.g. Twitter, email, certain blogs) and disable access for a certain length of time. You can't undo it until the timer runs out.
- Digital Detox: A free app for Android smartphones inspired by Digital Detox week.
- WriteRoom: A distraction-free way to write. Full screen, with just the text.
- iA writer: an app for Mac, iPad and iPhone that helps you focus on writing.
- StayFocused: A Mac app that enables you to block certain time-wasting sites, and choose how long per day you are allowed to spend on them.

- Readability: Removes clutter including adverts and icons from any web article, leaving just the content.

15 RECHARGE

If every waking second is filled with an information stream, when can the mind reset? Its power is being called on constantly. You wouldn't expect a battery to go on for ever; your brain is the same.

Everyone needs a break from what they are doing to recharge and have new ideas.

Why is it a good idea?

- Your body gets a break.
- Your mind attends to something else entirely.
- You'll end up feeling refreshed.
- You'll have new ideas when you get back to your task.
- Switching off regularly helps you avoid burnout.
- No one can be highly focused all the time – we're not robots.
- Focus is particularly efficient in short, intense bursts, so you need a break.

Recharging should be about getting out and doing something or learning something, or relaxing completely. Some ideas of how to switch off are by learning a foreign language or instrument, walking or running, doing yoga or dancing, painting or drawing.

We'll look at recharging in more depth in Chapter 11.

NINE STEPS TO FOCUS

Well done, you're almost half-way up the nine steps!

How are things going? Take a few minutes to note down your answers to the questions below in your notebook. Also think about anything you want to work on in the next chapter.

STEP FOUR

1. *How are you feeling now about digital distractions?*
2. *Have you noticed any resistance, for example, things you want to try but are putting off?*
3. *How are the people around you reacting to your changed digital distraction behaviour?*

A QUICK RECAP

- Would a deadline help you focus on getting through a task?
- What is your Most Important Task today? Spend some time and energy deciding on your priority each morning.
- Could you delegate your distraction so that someone else sifts through all the information?
- How many 'loops' do you have open at any one time?

- Is time tricking you as to how long you actually spend online?
- Are you expecting too much of yourself? Are you expecting yourself to respond faster than each message arrives? Slow down.
- If a goal seems overwhelming, break it down into chunks and work on these instead.
- Use tech tools to help you disconnect and focus on your work.
- Recharging helps with new ideas and gives you fresh energy.
- Willpower is the best tool of all.

FURTHER READING

- Another great productivity book by Leo Babauta is *The Power of Less*. Read this for tips on Most Important Tasks, batch tasking, and how to change your attitude towards email. Babauta, L. (2009) *The Power of Less: The 6 Essential Productivity Principles That Will Change Your Life*. London: Hay House.
- A productivity classic to read if you need a moment of inspiration is David Allen's *Getting Things Done*. Allen, D. (2001) *Getting Things Done*. London: Piatkus Books Ltd.

CHAPTER 8

Watch out for being caught out

When we divide our attention while trying to encode or retrieve memories, we do so about as well as if we were drunk or sleep deprived. MAGGIE JACKSON

BY NOW YOU SHOULD HAVE some idea of the things that are distracting you. In this chapter, we'll look at staying motivated to avoid distractions.

It's time to think about what you want out of a different way of doing things – day-to-day, at the end of the year, and in terms of results or relationships.

It's helpful to think on a bigger scale about these things so you have a powerful reason to resist the urge to check your email. Telling yourself: 'Don't check it just because you shouldn't' isn't always motivation enough.

Knowing why you want the time gives you the willpower you need, or makes your brain do that extra bit of thinking to organise getting on with something more important than mindlessly surfing the web.

It's not always easy to go against the crowd. You're pushing through a tide of people surging down a city street 'excuse me', 'excuse me', being pushed back again and again. Everyone else has their email on all the time... they give you a strange look when you say you haven't checked yours today. Everyone else is talking about what they heard on Facebook. But you've decided you don't want to spend time there any more.

You might feel alone on your new path, but you're not. A wave of dissent has started to build. Opinion formers have begun to ask difficult questions such as: Do we need to shut down? Are our relationships with smartphones unhealthy? Have we created a culture of distraction?[1]

Even for the converted, breaking free of digital distraction is a difficult and constant process. Digital distractions are waiting to tempt you every time you connect.

How then, can you avoid some classic traps?

SWITCH-TASKING

> Doing several things at once is a trick we play on ourselves, thinking we're getting more done. In reality, our productivity goes down by as much as 40 per cent, because we don't – and can't – multitask. We switch-task. Rapidly shifting from one thing to another, interrupting ourselves unproductively, losing time in the process. PETER BREGMAN[2]

Does switch-tasking sound familiar to you? You flit – super speedy – from one browser tab to the next, then you're back to email, then posting a quick Tweet, then back scanning web pages...

If you asked most people if they felt like they were going fast or slowly while doing this, the chances are they'd say fast. If you asked them if they thought they were using their time well, they'd say yes.

There's nothing wrong with trying to use your time well.

If you're anything like me, getting the most out of your time is really important to you. We need to remember that's all we're trying to do when we zoom off across seven digital platforms at once.

However, although it feels fast, and efficient, switch-tasking (or multitasking as we like to call it) actually slows you down and makes you sluggish. It's a classic trap.

When you switch-task, your productivity goes down by about 40 per cent (with some estimates showing that you lose as much as 75 per cent of the information you have just acquired).[3]

Why is this? Firstly, each time you switch, you have to read back in to what you were doing before. It takes time to remember what you were doing and find your place, and to forget about the task you switched to that is still on your mind. We can take 25 minutes to return to an interrupted task.[4]

Secondly, there is a high chance you will get completely distracted and meander off along an entirely different digital path, never getting back to what you were doing – especially if the task is difficult.[5]

How often do we face these interruptions?

One estimate suggests that workers switch tasks every three minutes during their working day, and that nearly half of those interruptions are ones they create themselves.[6]

The toll taken is not just on our time. Some estimates suggest that people distracted by email and phone calls suffer as much as a 10-point drop in IQ.[7] This is the equivalent of losing a night of sleep – a sluggish state most of us wouldn't opt for, given the choice.

Our memory suffers, too. Distraction expert Maggie Jackson is vocal about how serious the problem is: 'A culture of divided attention fuels more than perpetual searching for lost threads and loose ends. It stokes a culture of forgetting, the marker of a dark age.'[8]

What is actually happening in our brains?

When our attention is split, we are not encoding the information, and so don't form memories. There's nothing there to recall later. Have you ever experienced talking to someone who is saying 'hmm, yes' while doing something else, but later has no recollection of the conversation while you say, exasperated 'but I told you'? This is what has happened.

As James Borg explains, while your brain may have registered some experiences in its short-term (or working) memory, 'without concentration and full attention it never reaches the long-term memory stage. It's being asked to split its focus onto too many actions.'

'Multitasking is the enemy of memory,' he warns.[9]

FOCUS DAY-TO-DAY

Every day that we show up at the office, our desk, or our front room, we have to expect that people and things will try and distract us.

Emails will be clambering over one another to get to the top of the pile. Tweets will be shouting, trying to be noticed above the noise, Facebook updates will be luring us in to a merry-go round of nostalgia and fantasy. The Internet will be inviting us in, with a freshly painted Google front door.

How can we possibly say no?

Start by asking yourself these two simple questions:

Do you want more time in your day?

What do you want that time for?

Think big, we're talking major dreams and aims here. What is it that you really want the time to do?

Maybe you want to write a screenplay? Maybe, secretly, you'd love to move abroad? Maybe you've been dreaming about setting up a tea shop for years?

This goes to the core of what you are making the space and time in your life for.

While it may seem outlandish to be asking such big questions when we're talking about the time you can save in just one day (five minutes here, five minutes there), you have to trust that these minutes will multiply.

Don't be concerned about the scale of your plans in contrast to the scale of the day-to-day bits of time you are saving. You'll quickly notice – and be amazed at – how time saved on absence of distractions adds up.

If you've got more than one dream project, that's great. Once you've done one thing, you can move onto the next. Firstly, decide on the aim that motivates you most.

Got it? That's great. Fix an image of that new home, exotic holiday, or perfect job in your mind…

So where can you save the time to do it?

By cutting back on the distractions that tempt you day-to-day:

- Daily emails
- Internet
- Mobile phone/smartphone

- Reading updates on social media
- Posting updates on social media.

Which distractions have you most identified with in the previous chapters? This is a good place to start. Taking one particular distraction, ask yourself:

Where could you cut two minutes of distraction today?

And just to remind yourself:

Why do you want to do this?

Once you've planned your first small way to save time, add in a few more. Write down five ways that you could cut down distractions day-to-day:

1	
2	
3	
4	
5	

Remember, we're talking five minutes here, five minutes there – not a complete overhaul of how you live.

Once you have a list of five ways to cut distractions, try and do one of them every day this week.

If avoiding distractions day-to-day becomes a habit, you'll find focusing on results, long-term aims and relationships easier.

You'll be saving time by not switch-tasking. You'll be getting on with projects without allowing interruptions.

You'll start the day with productivity rather than with distraction. For a small change, it really pays.

FOCUS ON YEARLY AIMS

Many people tend to be good at setting yearly goals on 1 January.

But, when it comes to setting yearly goals at any other time of the year, we don't really think of doing so.

This is a shame, because, at certain points in our lives when we want to make a change, we really could do with a 'new year' to crystallise our aims.

Let's start a new year today.

By the end of this year, what do you want to have to show for a year of your life?

Hundreds of answered emails, a social media account brimming with updates, and a lengthy history of web pages visited? Or something more than that?

Sometimes, when we look back over a year, we get disappointed.

Have you ever got to the end of the calendar and realised that you got none of the major things that you wanted to get done, done? But yet… you've been really busy the whole time. What went wrong?

A classic symptom of digital distraction is that you feel busy all the time. So you feel like you are achieving things on a day-to-day basis, but, when you add things up at the end of the year, you haven't got much to show for 365 days of toil. No wonder it's disappointing. You've been busy, and tired, and haven't had a minute to yourself, and yet you still haven't got what you wanted from the year.

Remember that being constantly 'wired' makes you feel tired. But notice the difference – tiredness that comes from focused work does not have the same draining effect.

Being wired also means you are likely to be caught in another classic trap – never having a moment to yourself. Other people are always demanding your attention. Other people's demands are always directing your thoughts.

What could these minutes you never have to yourself have been used for, if you had got to keep them?

You need these minutes to improve your year. You need them to plan the things that will make this year look different. You need them to take steps towards the dreams and the goals that really – 12 months on – are your aims.

If we fill every spare moment with digital distractions, there's no time to sit down and actually book that dream holiday. In fact, there's not even time to have the inspiration to decide where to go.

By 31 December, you may not consciously equate all the time you spent on social media with the reason you didn't have a great holiday this year. But you left no time for long-form thinking. You left no time to plan, or to create.

Imagine that today is the first day of a new year for you (regardless of the month), and choose how you'd like the year ahead to look.

Five things I'd like to do this year:

1
2
3
4
5

Each time you are about to log in for a dose of digital distraction, take a second to hesitate, and ask yourself: Do I want to check my emails or achieve my big aims for the year?

Then spend, for example, five minutes learning the language you want to master, or half an hour picking out a new painting to brighten up your home.

By the end of your year, you'll be able to look back proudly.

FOCUS ON RESULTS

While some people are motivated by dreams, others are driven by results.

Particularly in the workplace, this is a useful area to focus on.

Changing your attitude to digital distraction could give you a competitive advantage, lower stress levels, and increased productivity.

If everyone around you is 'lost' in a confused and overloaded world of distraction, taking a step back from this will immediately set you apart. If you check your emails twice a day so that you can spend the time on doing a better job for the people you work for, it will quickly be noticed.

You'll also feel happier and more in control.

Leave things too long, however, and you risk the opposite – being the distracted one left behind.

Knowing you are the last one in the office to come up with a digital strategy that allows you to focus is not a good place to be. You'll be frazzled and desperately struggling to keep up, while others around you get ahead.

Don't worry though, you're already ahead (you're the one reading this book after all...).

Here's a quick four-step process to use in the office tomorrow:

1 Pinpoint your most productive time (look back at Chapter 5 for more details on this).
2 Reschedule your diary, if necessary, to free up one hour of this most productive time for actual work (meetings and calls should be swapped to less productive times).
3 Turn off all digital information streams during that time, for one hour (that includes email/smartphone/BlackBerry/RSS feed/any other feed/social media/Internet browsers).
4 Work for one hour. Allow no interruptions and be strict about this. Notice how much more than usual you achieve.

Setting aside one hour a day for the work you are meant to be doing should not be asking too much.

If you manage this, you'll soon see significant results. If you want, you can go further – try one hour in the morning and one hour in the afternoon.

For those keen to wholly embrace digital productivity, you can even go so far as to set up a no-digital island in your office (taking your lead from the neuroscientists in Utah who went on retreat, and the web designers, who we looked at in Chapter 6).

Many people are surprised at just how much they can achieve in such a simple way – by changing how they work for just one hour. They are also surprised at how fast it is to do this.

Swapping one highly productive hour for, say, three entirely unproductive hours makes sense. In working like this you also free up plenty of time.

How many businesses have actually thought through a strategy in terms of digital distraction? Not many, sadly. How many have just reverted to the default mode of always 'on'? All the rest.

Once you create even a simple strategy, you can sit back and watch productivity soar.

If you manage a team, you should be particularly alert to the impact a distracted worker can have on everyone else. A classic example of a distracted worker is a person who sends an email to the wrong person, or sends an attachment not attached.

Simple things can multiply the distraction factor in an office – such as when an employee copies in too many others (often for political reasons) to an email, or sends an email unthinkingly to an entire distribution list. This means up to hundreds of people are distracted, and switch tasks while they read that email (that may not even concern them). Productivity dips as they all lose time switching back to their original task. Think about whether an email needs to be sent at all. Add the simple line 'no need to reply' at the bottom of an email if there isn't.

Once you have mastered distraction avoidance techniques, pass them on to others around you.

FOCUS ON RELATIONSHIPS

'It's me or that damn phone.'

Ever heard anything along these lines? Then it's time to

look at how digital behaviour is changing your relationships before it's too late.

We need to look at three different relationships here:

1. The relationship in the room. These are the people we are talking to face-to-face (or rather not talking to because we're too busy looking at a screen).
2. The relationships at the other end of the line – the people reading our status updates or the people we are sending emails to.
3. The relationships we have with our phones and computers.

The relationships we have with our phones may seem a strange concept to discuss. But in his book, *The Fix*, Damian Thompson argues that more and more of us are developing addictive relationships with our smartphones.

In a recent article he talks about what he found when he was working on his book about addiction:[10]

> At Stanford University in California – just a stone's throw from Apple's headquarters – 44 per cent of students claim to be either very or totally addicted to their smartphones. Nine per cent admit to 'patting' them. Eight per cent recalled thinking that their iPods were 'jealous' of their iPhones.

These are strong feelings, and shocking findings. How can we feel like this about what is, essentially, a piece of machinery? Perhaps the machinery of our brains is not entirely under our own control...

'The makers of smartphones know far more about our mental reward circuits than is good for us,' says Thompson. He talks about how these devices 'practically force you to perform repetitive rituals of the sort associated with obsessive compulsive disorder'.

If you suspect you need to assess your own relationship with your phone, ask yourself these questions:

- *Do you get anxious if you forget your smartphone for a day?*
- *Do you sleep with your phone under your pillow?*
- *Do you ever switch your phone off?*

Be honest – what do you really feel about your phone?

Do you think of it at all as having human-like characteristics? Do you feel like it is there for you, or understands you? Do you let anyone else go near it?

Over the last few years, our smartphones have turned into very different objects from simple calling devices. Often, we feel they hold our lives.

Why do you like it and why did you get it? Because it makes you look good? To have the latest technology ahead of other people? To save yourself time?

Our relationships with technology are changing and becoming more complex. Being aware of how you see digital devices at this stage will help in the future.

Aidan Woods, an 18-year-old university student, calls himself an 'Apple fanboy'. He never switches off, and is one of the most connected among all his friends. To him, multitasking is a way of life.

Digital devices are a big part of my life. I have an iPhone5, a Macbook and an iPad, as well as an Xbox and a Playstation.

I want to keep up. I got the iPhone5 on the first day it was in the shops.

My phone lets me multitask easily, and the iPad is just the same – I just switch tabs. I multitask a lot. I'll go on Facebook while watching YouTube and texting at the same time, for example.

Let's see... I've got 24 applications open at once on my phone right now. It's about the same on my iPad when I'm on it. There are five tabs open on my computer.

Some of my friends are exactly the same. But a few don't use any devices. I'm probably one of the most extreme. I'm an Apple fanboy!

We communicate by texting – that's the most popular way. Or instant messaging. Sometimes we use phone calls.

I never switch off – I'm switched on all the time, even when I'm asleep. But I do always turn off the audible email notifications.

I never feel tired out by the need to be constantly connected. It's what everyone does. I like it. I'm not sure why.

I can remember life being another way. I got my first phone when I was 10 years old.

Why do I like being so connected? I can watch whatever I want, whenever I want on my computer. I can watch American TV. I can talk to someone on the other side of the city without having to go and get on a bus – talking on Facebook or texting is cheaper.

There are some downsides, like people not getting enough exercise. There are social issues, too. I know people who spend so much time on the Internet that they find it difficult to talk face-to-face with other people.

I have got into arguments in the past with my parents – for example, if one of them wants to talk to me and I'm on the phone, I can get into trouble for not answering. I also fell out with them about wanting to buy a new phone.

In the future, I think devices will get thinner, not necessarily smaller. There'll be a hologram screen – 3D without glasses. These things are already being developed. My guess is we'll have them by 2020.

IN THE ROOM

Some of the energy we put into relationships with our digital devices may be more usefully spent on the relationships in the room.

A classic example of where we are damaging these relationships is the BlackBerry on the restaurant table. There is clearly someone else muscling in here on an intimate dinner for two.

What does this signal to the people we are with?

We've fallen into a trap of thinking certain behaviour is acceptable when, in fact, it's downright rude. We've started to take our 'old' (face-to-face) relationships for granted and treat these relationships without the regard they deserve.

Let's take a step back and think about how these relationships might be suffering.

Think about your answers to these questions:

- Which relationships do you value?
- Do you use digital devices when you are with those people?
- Are there relationships in which you feel checking your smartphone in company is OK?
- Are there relationships where you make a special

effort to limit your use of your smartphone or laptop? If not, should there be?

- Have you ever been annoyed by someone checking their phone in your company? Who?

How many times have you seen two people walking down the street, both on their mobile phones? In many cases they are not even questioning their behaviour.

These are the things you are effectively saying each time you do this:

- There is something/someone more important than you at the other end of the phone/computer.
- I don't want your (human) company.
- I don't value our time together.

Next time someone checks a smartphone while you are with them – breaking off your conversation – notice how it makes you feel. Next time you are with someone else and you check your phone, ask yourself how they might feel.

When a phone is visible, it often makes us think of wider social networks. And there is evidence that this is damaging our face-to-face conversations.

Research from psychologists at the University of Essex showed that when a mobile is visible during a conversation – even when it is not being used – people feel less close to the person they are talking to.

The mere presence of a mobile had negative effects on closeness, connection and conversation quality, and people were less likely to feel the person they were talking to showed empathy and understanding. If something personally meaningful was being discussed, these effects were even more pronounced.[11]

So, if you want to have a serious conversation, make sure your mobile is out of sight (or, even better, switched off).

But how do you get the person you are with to check their phone less? There is often tension when it comes to setting boundaries.

Things you might say include:

- I'm not meeting you for lunch unless you don't check your phone.
- No phones at the dinner table.
- No computer after 9 pm.

In many of these cases, the person laying down the boundary is seen as the unreasonable one, the nagging one. Yet they may be doing their loved ones (as well as their relationship) a favour.

Having to fight to get attention doesn't make you feel valued. Is 100 per cent attention too much to ask these days?

Spending time with someone and then ignoring them means you're together, yet apart. What you are losing is quality time.

OUR NEW BEST 'FRIENDS'

While we have been quick to shun our old face-to-face friends, we have embraced enthusiastically a host of new 'friends', many of whom we don't even know.[12]

Social media has redefined the very notion of a friend.

But, if we look closely, we may find that these online relationships are taking up an inordinate amount of our time.

Relationships in this sphere are fraught with issues – of online persona, of reading signals, and of keeping up with the Joneses. How we want to appear to others takes up much of our time on social platforms. Social construct and identity are two major areas of study in this field.

One thing that keeps us hooked on social media is fear of missing out (FOMO), as discussed earlier. It makes us check, check again, and check again. Something might have happened! Someone might want to talk to us! There might be a new message there!

We arrive at social platforms driven by fear (of missing out). We often leave feeling inadequate, left out, or overwhelmed. It's not exactly an advert for happiness.

In many cases we are spending so much time with these new 'friends' that we are forgetting cues for social interaction. It's hard to tell the depths or precise nature of what someone is really thinking when all you see are words on a screen and the occasional smiley ☺ or sad ☹ face.

There is speculation that technology is reducing our ability to show empathy, as we meet up in person less and less.[13]

The distinction between the public and the private is another area where the norms have changed. There have been many 'Tweet and regret it' mishaps.[14] But, even for those who haven't conducted a high-profile break-up in public, getting used to things being said in public that used to be private is something that we are still coming to terms with.

The chance to see in to other people's worlds is what draws us in to social platforms.

The distraction we face from other people is something we will discuss more in the next chapter.

NINE STEPS TO FOCUS

Brilliant! You've passed the half-way mark as you take a big leap up to step five. Can you remember how different things looked as you stood at the bottom of the steps? Let's see how far you've come.

STEP FIVE

1. *What is working well for you in tackling digital distractions?*
2. *Which new behaviours are now part of your everyday life?*
3. *What else have you noticed changing in your life?*

A QUICK RECAP

- Switch-tasking feels efficient but you cut your productivity by 40 per cent when you do this.
- Saving five minutes on distractions day-to-day is a way to create time for the projects you've always wanted to do.
- Look back at the end of this year and be proud of what you've achieved by making space to think and plan.
- Don't be the one left behind at the office. Improve your results with a simple four-step plan.

- There is evidence to suggest we are developing unhealthy relationships with our smartphones.
- Face-to-face relationships are suffering due to digital distractions. Beware of giving the signal 'there is something more important than you'.
- The new relationships we have created online are sometimes driven by fear.
- Reading emotion is becoming an issue for those spending a lot of time conducting relationships online.

FURTHER READING

- Maggie Jackson makes no pretence that everything will turn out OK in her book, *Distracted*. She backs up her arguments with scientific research, and predicts the coming of a dark age. Jackson, M. (2008) *Distracted: The Erosion of Attention and the Coming of the Dark Age*. New York: Prometheus Books.
- Cameron Marlow is a research scientist and 'in-house sociologist' at Facebook. Read some interesting papers from him at http://cameronmarlow.com/.

CHAPTER 9

Who's distracting you?

What information consumes is rather obvious: it consumes the attention of its recipients. Hence, a wealth of information creates a poverty of attention... TIMOTHY FERRISS

IF THOSE AROUND YOU ARE distracted, you can easily slip back into distracted behaviour yourself.

Let's start with a quick question.

Who is distracting you?

You may well have one person in mind who you really wish would change their digital habits. You may feel that their digital behaviour is not just affecting their life, but your life, too.

You want to be screen-free, and spend time together, but they continue to stare only at their smartphone. So you give in, and pick up your phone, too...

In this chapter we'll look at diplomatic tactics you can use with the person in your life who is digitally distracted. We'll also carry on from where we left off in the last chapter and look more at relationships.

Sometimes, we are quick to blame other people for distracting us. We are also quick to blame technologies that offer us a distraction.

But are these things just a convenient excuse for our own behaviour? Way before the Internet was invented we knew how to put things off. So, while we need to look at distracting forces around us, we also need to keep focused on our own aims – which give us the power to resist distractions.

Maybe, as well as noticing that certain people distract you, you'll have a niggling feeling when it comes to relationships overall that something has changed. If that's the case, you're right.

Despite being connected all the time, we're finding it harder to communicate. It almost seems like we've forgotten how to.[1]

No wonder we're struggling. Plain English sitting in the front room with a cup of tea used to be enough. Now we're operating on a global scale, sending abbreviated messages littered with symbols to people we don't even know. The old timescale? First-class post if it really mattered. The new timescale? Faster than instant if we could. We'll look at how this is changing the way we relate to one another.

WINDOW SEAT

Do you remember how annoying it used to be when people used the classic opening line on their mobile: 'I'm on the train…'? It's still just as annoying. But have you noticed that it happens less and less? We don't talk any more. We text. We stroke screens. We IM.[2]

The train is a great place to observe how we, as a society, are changing the way we use digital devices. And, by observing how other people are behaving, we can often get an insight into our own ways of doing things.

Imagine you're on a train. There are two people opposite you. One is reading a book. One is reading a screen on a mobile device.

You watch them, and wonder: who is happier? Who is more stressed? Who is more likely to have an original thought?

These are tricky questions to answer without knowing more about these people (happiness depends on many factors, after all). But, in asking these questions, we are starting to observe the behaviour of others and question it.

Travel is a classic example of 'gap time' that we could use for thinking and resting, that we now tend to fill with digital distraction.

It infuriates me when I look around a train carriage and see person after person after person transfixed by their mobiles, heads bowed. What could be so important?

Outside the window there is so much to inspire. But we're missing out, and so our landscape has changed.

Perfect sunset? It doesn't matter. The rubble of the 357th email of the day is holding our gaze.

FIND A GOOD EXAMPLE

As well as noticing bad examples, we can learn from the good examples we spot around us.

Take a moment to think of the most focused person you know.

How do they act? What things do they do that make them efficient? How do they appear in terms of temperament?

Write down three things about their behaviour that you notice:
1
2
3

Now, think of the most digitally distracted person you know.

How do they act? How do you know they're digitally distracted? How do they appear in terms of temperament?

Do they seem happy, relaxed and productive? Or do they seem stressed and overwhelmed, and never have time for you?

Again, list three observations:
1
2
3

Who gets most done out of these two?

What do these two people do when you text them? Call them? Post a status update?

Now, line up the five people you spend the most time with, from most digitally distracted person, to least digitally distracted.

Where do you come in this line?

Are you more distracted than most of your friends and family, less distracted, or about the same?

Don't be surprised if the people you spend the most time with tend to have similar levels of digital distraction to you.

We tend to be like the people we spend most time with.

Want to be richer? Spend more time in circles where people have more money. Want to be more focused? Spend more time with people who are not distracted.

If everyone around you is distracted by digital devices, beware of thinking this is the norm.

FAMILY TIME

> We have found ways of spending more time with friends and family in which we hardly give them any attention at all. SHERRY TURKLE[3]

What if the people distracting you are in your own front room?

The norm within some families has become being together, connected to screens, yet disconnected from one another.

> For all our constant connectivity, our electronic devices often keep us apart. ELIZABETH BERNSTEIN

A family might, perhaps, all sit at the same table, staring at different screens (mobile devices, tablets, laptops), not talking to one another. Mother is checking her BlackBerry work emails while father is on Twitter following a stream of updates. Son is IMing and daughter is texting friends. Or they each go off into a separate room, to stare at a separate screen.

This isn't limited to the living room. I watched a family do this recently at a swanky restaurant in London – the kind of place you go for a very special occasion (to stare at your screens…).

It certainly redefines family time.

Bernstein charts examples of families that have noticed the damage they were doing to their relationships, and decided to introduce a tech cleanse.[4] Be warned, she says, it's not easy.

A technology cleanse, or tech detox, is where you – as a family – unplug for a short time for the good of the relationship.

One great account of a family that went through a tech cleanse is *The Winter of our Disconnect* by Susan Maushart.[5] We'll look at this in more detail in the next chapter.

Tips for families who want to try a tech cleanse include telling your family in advance so they can get used to the idea, not just swapping technology for another isolating activity, and weaning yourself gradually off gadgets.

A tech cleanse could also usefully be done in the office, or among friends. Make sure you have good activity to replace being connected so you don't get bored (such as cooking a great meal together).

Louise Ladbroke is distracted by the distraction of those around her.

'Tea's ready,' I say. My partner is engrossed in something on his computer. My son is avidly playing a game on the iPad. No response.

If I disappeared, I think, how long would it be before they noticed? Half an hour? An hour? Maybe two?

Sometimes I make use of this. If I want to do something, and I don't want my seven-year-old demanding my attention, I know the iPad will buy me half an hour before he even looks up. But, somehow, somewhere, I also know that's not right.

When we do talk, it's no better. My partner had a 10-minute conversation with me the other day about ether ports: where mine were, and whether they were working or not. I managed to sustain it, having no idea what, or where, they were.

When I talk to my son, I try to tell him about how the Romans used to build houses or go to the toilet, for example. He pays little attention. But, if this information is coming from a computer, it's like the holy grail. He listens avidly, obeying instructions like a hypnotised member of a technology cult; the God that is computer is more holy than the human. If only it could get him to tidy his room…

On a French airline once, I remember looking at the in-flight entertainment. They called it the 'Programme de Distraction'. It made me laugh because it was so direct. When I was at school, to be distracted was a bad thing. Now it seems it's a human expectation, a need to get us through those vast terrifying expanses of time where nothing much might be happening.

We fill this void by always being available to others, on-call if you like, through texts, emails, and even tweets and Facebook. If we don't address these things, we're lagging behind, neglecting our friendships, not being conscientious enough with our business. It ups the stress levels for me, for one, as I now just have more and more social interaction to keep up with.

Sometimes I feel desperate to live in the moment. I can be having a really great evening out, or even a quiet one at home, and my phone will flash up that I've got an email, or a text, and I'll read it. It can often hijack my mood for the rest of the evening.

We assume all technology must be good. It is, but only in moderation. Like a drug, it can enliven a situation, but it's craved more and more until it's over-used and out of control.

Technology is supposedly for our convenience but, for every tiny, time- and labour-saving device out there, it feels like there's a karmic payment somewhere else, a room jammed full of chargers and leads, or computers containing the information you can upload onto your device.

The same karmic payment goes for time. For every great evening I might have with my partner going for a nice meal, there's an evening lost somewhere else where we spent hours surfing the Internet to find the right restaurant. That's technology Karma.

LEARNING THE HARD WAY

What do you do if your teenager, or your parent for that matter, seems to be literally on a different digital planet to you?

When looking at the digital habits of someone of a different generation, we need to appreciate the world they grew up in. Many young people have grown up knowing nothing other than being connected, and it has had a great impact on their behaviour, communication and relationships.

Sherry Turkle explores the behaviour of young people, and discusses how relationships (for young and old alike) have changed in a connected world.

She notes that, before mobile phones become an essential element in a child's own life, they know them as the competition:[6]

> Children have always competed for their parents' attention, but this generation has experienced something new.

Turkle says that what children long for is 'full attention, coveted and rare'.

It makes it worse because the parent is there, yet not there:

> Today, children contend with parents who are physically close, tantalizingly so, but mentally elsewhere.

Turkle gives the example of a son and father sitting next to each other on the couch watching a football match. The father is also on his BlackBerry. Precisely because they are so physically close, the son feels like his father giving attention to the BlackBerry, not him, is particularly excluding.

Turkle points out that we rarely have each other's full attention, despite constant connection. It is a depressing sentiment. It brings us back to the notion of attention as the most precious commodity of our time. We haven't understood this yet, and are suffering the painful consequences of giving our attention out to everyone everywhere, at the same time as not fully giving it to anyone any more.

Babies waving wildly in the pushchair, and toddlers pointing out 'look, look' while a parent pushing them looks at their phone, just want the same thing – attention. But, if the attention goes to the smartphone, the child has to up the ante and shout louder to be heard.

Once these children grow up, and get their own mobiles, split-attention is the pattern they know.

> Although everyone multitasks, the Net Generation and the iGeneration perfected it to a science, a way of life. LARRY ROSEN[7]

As we have seen, teenagers are consuming 10 hours and 45 minutes of media per day, once multitasking is accounted for – spending more time on this than many adults do on their working week (and often teenagers are using media seven days a week).

This estimate doesn't even include time spent texting. Research from the Kaiser Family Foundation shows that media use among young people keeps rising (the study was repeated in 1999, 2004 and 2009). It also shows that, if parents set any kind of rule about media consumption, young people will consume almost three hours less media per day than those whose parents set no rules.[8]

Here are two questions to consider:

Could the people whose digital distraction habits annoy you actually want attention?

When do you give your attention to a digital device instead of a person when you shouldn't?

IN THE OFFICE

The typical office breeds distraction. For a start, it is full of inviting computer screens. Then there are colleagues, wanting an instant email response, insisting on using IM, or interrupting you with questions.

There's the boss who presents you with a BlackBerry ('A gift!') and the assumption you have become an on-call worker. No pay rise, just a shiny new BlackBerry in a box.

Expectations to be constantly connected in a work environment are dangerously high. This constant connectivity goes against much established productivity thinking.

In his book, *The 4-Hour Workweek*, Timothy Ferriss says we should focus on being productive instead of busy.[9] Being busy is often used as a guise for avoiding the few critically important but uncomfortable actions we need to take, he says. The options are 'almost limitless' for creating busyness in an office. Ferriss gives an example:

> ... I'd love to talk about the gaping void I feel in my life, the hopelessness that hits me like a punch in the eye every time I start my computer in the morning, but I have so much work to do! I've got at least three hours of unimportant email to reply to before calling the prospects who said 'no' yesterday. Gotta run!

He says the first step is to 'develop and maintain a low-information diet'.

To keep your own stress levels down and productivity high in an office, the first thing to notice is who is most distracting.

Distracted people do these things:

- Bring the phone to the table.
- Only half-listen to you.
- Always try and do two things at once (watch TV while online shopping).
- Make mistakes due to digital distraction.
- Blame their 'bad memory' for forgetting things.
- Take longer to complete tasks.
- Defer immediately to their smartphone or the ping of an email.

- Talk about things/people/stories learnt from social media not face-to-face.

How to spot focused people:

- They produce great results.
- They are the ones having the ideas.
- They produce a high volume of work.
- They get irritated by unnecessary distractions.
- They look like they are getting on with their work.
- They look absorbed not bored.
- They remember things.
- They leave work on time.

Avoiding distractions in the office means you will be able to be absorbed in a task and produce high-level work and great ideas.

Most managers want this for their company and their staff, yet they need to promote the right culture and lead by example.

It's a simple thing but, if you expect your workers to reply instantly to an email, it means you are expecting them to live with their inbox always open in case an email arrives. This drains productivity and wastes attention.

One estimate from the University of Toronto suggests some 12 per cent of an average company payroll is soaked up by the unproductive use of work email.[10]

I'd say this estimate is a bit low for most offices. Take a walk round your office tomorrow (especially if you're the manager there) and guess at the percentage of people who have email open on their screens. 40 per cent? 60 per cent?

Switching off email for an hour is the easiest instant way to boost productivity. Try it. The second time you try it, switch off social media, too.

Encourage focus in email and written documents. This means short emails, and only emailing when necessary. It means switching off email whenever you can, if needs be, explaining you are doing so because you are working.

HOMEWORKING

There may be no colleagues physically present to loiter by your desk and tug at your shirtsleeves. But think that means no distractions? Think again.

Working from home brings many non-digital distractions (who knew emptying the bins was quite so high up the To Do list?).

But, as well as trying to ignore all the things that need doing round the house, homeworkers must turn a blind eye to digital distractions, too.

The very thing that means you should be getting loads of work done (there's no one to distract you) leads you to seek out distractions (you want some company). Social media can seem like the perfect solution.

Twitter and Facebook provide 'virtual colleagues' at the click of a mouse. As homeworker Angela Clarke says in Chapter 5:

> Like an office worker may turn to their colleague at the next desk and share their thoughts, so I turn to Twitter. I post about my frustrations, my happiness, and my observations on the pitfalls of eating a sugary doughnut over your keyboard (v. crunchy). And I do it all while wearing pyjamas.

She calls Twitter 'the reassuring background chatter to my day'.

But, while social media give homeworkers a feeling of having someone there, they can take up vast swathes of their time that they then have to make up for at the end of the day.

Decide how much social contact you need. Then think about ways to get that, other than just leaving your email and social media switched on all day.

Experiment with having two social online hours a day – for example, between 12 pm and 1 pm and 4 pm and 5 pm.

Look around, too, for other ways to connect with people throughout the day – have a five-minute chat with a neighbour rather than rushing by. Make a phone call to sort a task out instead of exchanging a five-email chain.

NO BOUNDARIES

The blurring of boundaries is something homeworkers and office workers alike are struggling with.

As one 21-year-old social media manager said to me recently: 'There's no start and end to it. I'm not really sure what's work and what's not.'

For homeworkers, though, there is no fixed start or end to the office day, either, and no distinction between workplace and home. This makes boundaries even more difficult to establish. While the office worker may imagine the homeworker is working fewer hours than them, the reality can be very different.

Expectation is a big issue for homeworkers. Are you putting pressure on yourself to be constantly available (to clients or to the office, if you are working remotely)? If you are constantly connected just in case someone needs you, you leave yourself open to being constantly distracted. So everything takes longer and work creeps into the evening.

Instead, make the most of a great advantage of being a homeworker – there's no boss. Work in focused bursts, don't force yourself to work through slump times, and avoid distractions. Results are the only thing that matter. If your work takes you three hours rather than eight, you win.

Some tips for homeworkers:

- Can you call instead of email? A human voice establishes closer contact, and one call is often more efficient than a five-email chain.
- Appreciate non-digital social contact – like a five-minute chat with a neighbour.
- Beware of other people thinking you don't have work to do because you are working from home (and expecting long chatty email exchanges, for example).
- Beware of thinking the same yourself.
- If other people expect you to be available, set predictable time off (see Chapter 6 for more on this).
- Use homeworking to your advantage and don't work through your slump times.
- Have some time away from the screen – ideally outdoors – every day.
- Put aside some social online tasks for the end of the day, as a way to wind down, rather than blurring work and play all day.

HOW TO STOP DISTRACTIONS FROM OTHERS

Identifying the people who are distracting you is one thing. Getting them to change their behaviour is quite another.

How to ask someone to pay attention to you, not their phone

The scenario:
The person you are with answers their phone or texts others when with you.

The problem:
You feel second best. They think it is normal behaviour.

What you would like to do:
Grab their phone out of their hand and switch it off.

Practical solution:
Decide to set aside 10 minutes of phone time every hour to both catch up on missed calls/texts/messages. Say you are turning off your phone so you can give them your full attention. It might get them to think twice before using theirs.

How to ask someone to write you shorter emails

The scenario:
This person always sends you emails that are like a badly written essay.

The problem:
Your attention and time are being wasted wading through searching for key points.

What you would like to do:
Not read their email.

Practical solution:
Ask them to summarise their email (in bullet points if they want to). Explain it is a lot longer than most emails. Say you will respond to their summary.

How to manage expectation on email

The scenario:
Other people expect a certain availability from you on email.

The problem:
Being always logged in to email is wasting your time and giving you a headache.

What you would like to do:
Never check your email again.

The practical solution:
A great email signature. Say that you're busy or working, and that you check email twice a day (for example). Once you manage expectations and state when you will be available, people won't worry if they don't get an instant response.

How to reply to an email that is like an old-style letter

The scenario:
A long email you receive (often from friends or family) has been crafted with care.

The problem:
You feel obliged to reply in a similar vein. You don't want to damage the relationship.

What you would like to do:
Write a one-line response. You're busy!

The practical solution:
Don't try and reply to this email amid the jetsam and flotsam of the digital day. Put it to one side until a quiet time, like the end of the day. Write using tone, anecdotes, and pace. Particularly with far-away friends this is one way the relationship stays alive.

How to tell your followers you won't be posting 24/7

The scenario:
You've built up a following on social media and now you feel like you should be there for your followers round the clock.

The problem:
You never get a break. There are always updates to read and posts to write.

What you would like to do:
Make it all go away and breathe.

The practical solution:
Be realistic. You should not be available 24/7. Don't try

and read all updates. Schedule some of your updates. Dip in and out and try and get back the sense of fun you had in social media before it was an obligation.

How to get someone to turn their phone off in a meeting

The scenario:
The meeting is interrupted by phone calls, or people typing on their smartphones.

The problem:
Without people's full attention, the level of discussion suffers. They also don't remember points from the meeting.

What you would like to do:
Ask them to stop being so rude.

The practical solution:
At the start of the meeting, ask people if they are willing to focus on this meeting, so that you get better results. If they say yes, tell them they will all need to turn off their phones for an hour to do that (or suggest 30 minutes then a phones-on break).

NINE STEPS TO FOCUS

Well done! You've got to step six. Hopefully a focused way of thinking is starting to pay off!

Let's check on your progress.

STEP SIX

1. *Have you noticed other people trying to knock you off course at all?*
2. *What would you like to try at this stage (don't be afraid to think big!)?*
3. *Is there anything you are still having real problems with?*

A QUICK RECAP

- What can you observe about digital distraction on the street, in the office, or on public transport?
- Think of the most distracted person you know. Now think of the least distracted. Where do you fall along this spectrum?
- Look at who you spend time with and their distraction habits. We tend to be like the people we spend the most time with.
- Could your family try a tech detox and use the time to reconnect?
- For young people who have grown up in a connected world, giving attention to just one thing can seem an alien concept.
- Your energy and your time are drained by digital distractions in the office. Find out how to spot them and change the way you work.

- Homeworkers are particularly susceptible to digital distraction because of the need for social contact.
- There are ways you can ask someone to write shorter emails, listen to you, or wait for a response on email or social media.

FURTHER READING

- For statistics galore on how we communicate and behave using digital media, see Ofcom's 2011 Communications Market Report.
- Sherry Turkle's *Alone Together* is really worth reading on this topic. Turkle, S. (2011) *Alone Together: Why We Expect More From Technology and Less From Each Other*. New York: Basic Books.

CHAPTER 10

Keep your aims in mind, and aim high

There is more to life than increasing its speed. MAHATMA GANDHI

It is indeed possible to live without Facebook. In fact, it's not only possible, it's (gasp) actually kind of rewarding. JESSICA TAYLOR[1]

CALL IT WHAT YOU LIKE: tech cleanse, digital detox, personal reboot, tech break or just plain old switching off, once you try it, you'll know it's good for you.

Sometimes, your first tech cleanse might not be entirely voluntary – perhaps you suddenly find you have no mobile reception, and break out in a cold sweat.

This is what happened to BBC producer, Stuart Hughes, who was forced into an unplanned digital detox in the wilds of Ontario.[2]

At first, he felt desperation at not being able to connect. But this gradually shifted to immersion in the real world – cheering on his son as he rode a horse for the first time and 'devouring' books by torchlight.

When Hughes switched back on he 'felt a wave of disappointment' as he checked Facebook and Twitter. His emails were 'irrelevant or out of date'. Twitter was 'as catty, witty, but ultimately transitory as ever'.

> I, meanwhile, had stored up timeless memories, perhaps made all the more precious by the fact they were personal and private, not tweeted, texted or emailed with my social network.

A story like this about what other people feel like when they switch off can be useful to hear if you're thinking of taking your own tech break. There are many similar accounts of 'social de-networking' on the web.[3]

Writing for *The Daily Muse*, Jessica Taylor talked about what she felt after quitting Facebook:[4]

> I started to realise that much of what we were communicating was trivial, and that, frankly, I didn't care about it.
>
> This website was sucking away more time than I devoted to many other, far more worthy areas of my life. When I realised I was spending more time on Facebook than I was on volunteer projects or even reading, I knew a change was probably in order.

You'll start to see you are not the only one going through this tricky process. Ask people around you – what do they feel about their own digital habits?

ON YOUR MARKS...

Let's be realistic – saying no to digital distractions time after time is not going to be easy.

The bad news? Evidence shows that media is harder to resist – even once we have decided not to use it – than many other temptations. We tend to fail in our attempts more than we do with many other things (the urge to use media is harder to resist than the urge to drink alcohol, smoke, or sleep).[5]

We really do need to use every trick at our disposal, and gather as much resolve as we can.

Don't be overly harsh on yourself if switching off is disastrous the first time you try.

Why is it so difficult to do?

We know why digital distractions are such a pull – because we have established new pathways in the brain and new habits, and because each digital distraction offers the tempting promise of an instant reward, and instant gratification.

> We may appear to be choosing to use this technology, but in fact we are being dragged to it by the potential of short-term rewards. Every ping could be a social, sexual, or professional opportunity, and we get a mini-reward, a squirt of dopamine, for answering the bell. TONY DOKOUPIL[6]

Receiving these 'pings' might start out with a feel-good dopamine high, but this can quickly turn to a low.

Our dopamine levels can plummet if we anticipate gratification that doesn't happen (for example waiting for a phone call or expecting an email that doesn't appear). The 'non-reward' is unbearable.[7] Avoiding dopamine rollercoasters is another good reason to turn your phone or email off so that you are not waiting for and anticipating a message. When you switch back on, on your terms, you'll see what's arrived.

Brain expert Gary Small says that, even if you are not addicted to the Internet, you may be 'struggling with its enticement'. He explains what happens with an Internet 'high' and why we keep chasing the rush.

'Self-proclaimed Internet addicts report feeling a pleasurable mood burst or "rush" from simply booting up their computer...'.

The feelings of euphoria are linked to brain chemical changes, says Small.[8] The system that controls these responses involves the neurotransmitter dopamine, 'a brain messenger that modulates all sorts of activities involving reward, feeling good, exploration and punishment'.

Dopamine transmits messages to the brain's pleasure centre, causing addicts to want to repeat those actions over and over again – even if the addict is no longer experiencing the original pleasure and is aware of negative consequences.

So with digital distractions, what started out as a 'rush', 'a new message!', can turn into a negative consequence: 'So many messages I'm overwhelmed...'. But, yet, we keep checking partly because of the original high.

If we feel strong resistance when we attempt to stop using a digital device we feel almost addicted to, we can now understand why.

FIVE WAYS TO BUILD UP FOCUS STAMINA

Ready to take a tech break? Let's go.

1 Don't try and run a marathon before you're ready

If you set unrealistic expectations and try and focus for, say, three hours straight on your first go, you'll be disappointed when it doesn't work out and will be more likely to give up altogether. Be realistic and take baby steps. Start with 15 minutes of pure focus and work up from that, adding on an extra 15 minutes each time.

2 Pick a role model whose style of focus you admire

Study that person. Learn all you can about how they manage to do all that they have achieved. Chances are, they're not wasting much of their time on social media (they've probably got someone else doing that for them). Now try and copy some of their methods. You might find out, for example, that they get their best work done in the mornings. Does that work for you?

3 Start with tasks you enjoy

It's easier to have pure focus when you're doing something you enjoy. So do those things you enjoy and get lost in them. If you find you're doing lots of tasks you don't enjoy, cut these out or down, or add in elements you like.

4 Set goals

Nothing focuses the mind quite like a deadline. To increase focus along the way, set yourself mini-deadlines and rewards all along the route: *'At 11 o'clock, I'll have a 10-minute break; at lunchtime, I'll meet up with a friend...'*

5 Make it easy for yourself

Remember those prime times of day for you? Focus during those times. Not everything can be done at a 'prime time', so make sure you select carefully what you let in to these slots of time. Once you can focus then, widen it out and start focusing at other times of day.

PUSH YOURSELF

The aim with a tech cleanse is simply to put you in control of how you use the technology, rather than the technology being in control of you.

Ultimately, you should be able to pick and choose when and how you use different platforms – you should set the rules.

Push yourself a bit. Depending on where you are starting from, this will mean different things.

These are examples of what to try at either end of the digital distraction scale.

Text-aholic

Are you welded to your phone? Try switching it off and putting it in a different room for 15 minutes.

Email-aholic

Do you always have email open on your screen? Experiment with switching off for 10 minutes, switching on for 10 minutes.

Social media-aholic

Do you post everything you do as a status update? Try keeping three things to yourself today, and telling friends in person instead of online.

At the other end of the scale, if you have already got to the point of balancing out your digital habits, take things a step further:

Mobile-lite

You can already function with your mobile switched off. Push it and don't take it with you when you go out. Switch it off rather than on as a default for a day, switching on now and again.

Email-lite

When you switch on your Out of Office you mean it – you are leaving the office behind. Go a step further and don't check your emails for an entire week.

Social media-lite

You don't define yourself by your status updates. If you've had enough of a social platform you don't use much any more, why not quit?

One organisation that pushed the boundaries further than you'd think they'd go is The Constellation. Here, global facilitator Olivia Munoru describes how this virtual office-based group switched off together for days.

Usually when I go to meetings, I find that people check their phones and jump on their laptops at every spare moment. Not this time. By choice we had no Internet. It made a remarkable difference. Especially because our meeting lasted four and a half days…

Arriving from different countries across the world, we met in a house more than 200 years old, in a small village called Houffalize, in Belgium. We were surrounded by hills and old farmhouses, a perfect setting for creative thinking and productive conversations.

Yet before this meeting, some of the team had never even met face-to-face. This is because we live in different countries.

Our day-to-day work is carried out through a virtual office, using applications such as Skype, Email, Dropbox, Google Drive and Google Sites. Together, we successfully run a global NGO (non-governmental organisation) with thousands of members across 30 countries. We create and manage partnerships with UN agencies, other NGOs, and businesses. We connect communities and link individuals so that they can share experiences and learn from each other.

The Internet is an essential platform for us to operate. But for these few days, it was non-existent.

Each day started with an exercise session, each of us taking turns to lead the group. One morning it was a walk through the farms, another it was yoga. The only interruption was a neighbour dropping by to offer us a bucket of milk fresh from the cow.

Each participant was responsible for organising a meal. We had home-cooked Indian, Indonesian, Dutch and Belgian food. The British participants served a ploughman's lunch followed by shortbread, strawberries and cream. I was responsible for the Aussie lunch... meat pies and beer anyone?

Each meal was cooked and served with love, enabling team members to bond as they prepared the food together. When we sat down to eat, in that beautiful old house with the exposed beams and old furniture, it really did feel like a family gathering.

In the afternoons we had time for 'walk and talk'. It was an opportunity for smaller groups to explore side conversations. This 90-minute session was around that time of day when people suffer post-lunch-lethargy. A very effective use of time.

This was by far the most productive meeting I have ever been part of. Our choice to go offline for nearly five days was, I believe, a big reason for this success. We were completely focused on each other and the tasks we had set ourselves.

Find out more about The Constellation at www.communitylifecompetence.org.

TAKE STOCK

After you've experimented with a tech cleanse, think back over how it made you feel.

Write down anything you noticed about it here – anything that happened (good or bad) or how you felt:

1
2
3

Did you feel tempted to give up on your experiment? Did you give in to that first wave of temptation to log in? It doesn't matter if you did – especially on a first try – the important thing to notice is what happened.

Did anyone or anything else distract you? (We have seen how distracting other people can be.)

Note what stopped you here.

These things stopped me:
1
2
3

What do you think you might try next time (for example switching off a different medium, switching off for longer, switching off at a regular time each week)?

A VITAL HOLIDAY INGREDIENT

A great time to experiment with switching off is when you're on holiday. You're already relaxed, and there are plenty of other experiences to make the most of instead of staring at a screen.

But worrying evidence shows that last year many people arrived back at their desks at the end of the summer having not had a break from work emails and social media.[9]

If you need help to get or stay motivated, look for a 'digital detox' holiday. Tour operators are starting to develop this kind of break as a new health holiday trend.

These holidays typically include no Internet, no mobile and perhaps even advice on how to switch off.

On digital detox breaks to St Vincent and the Grenadines, they are asking travellers to leave their technology at home. Other places might, for example, lock your laptop and smartphone in their safe when you check in.[10]

HOW TO TAKE A DIGITAL HOLIDAY

If you missed the chance to take a digital holiday when you were away this year, why not take one back at home.

A day of digital holiday could be just the thing to help you recharge.

Things you should switch off on your digital holiday:

- iPhone
- iPad
- Smartphone
- BlackBerry
- Mobile
- Laptop
- Computer
- Any other mobile devices.

If you want to go further, you could also include TV, newspapers and magazines (they are all full of information streams the brain has to work hard to process).

What you can do:

- Talk
- Read books
- Create
- Relax
- Exercise
- Sleep
- Walk
- Cook
- Play instruments
- Explore
- Plan a holiday (without using the Internet)
- Visit somewhere new

- Have ideas
- Write.

ONE CLICK AWAY

If you feel yourself wanting to log back in (on a digital holiday or day-to-day), try this trick.

Next time you are tempted by a digital distraction, stop yourself when you are one click further away.

This is an easier point to stop yourself at than at the log-in page or homepage, or once you are already logged in (think how automatically and quickly you type your password, then there's no turning back).

So, even before you click on the Twitter or Facebook icon on your smartphone, or before you go to the Internet shortcut on your laptop, ask yourself: do you really want to spend time on digital distractions right now?

Is there something else you would rather be doing instead, if you think about it? Is there something you would rather be doing to build towards your goals?

Ask yourself each time you're one click away – do I really want this distraction right now?

If the answer is no, close down the Internet browser, put down your phone, shut down the computer or disconnect the Internet if needs be. Then get on with what you want to be doing.

KEEP YOUR AIMS IN MIND

Remember the yearly goals that you set yourself a couple of chapters ago?

When things get really tricky trying to combat digital distractions, reminding yourself of these helps.

What is it that you want to have achieved in 12 months' time?

Look back at the list you made of five things you would like to do this year.

Are these still the same as they were?

Take a few moments now to think about your dreams.

What is it that you want the extra headspace you get from cutting down on digital distractions for?

If you've thought of anything new, or if things have changed, note down your answers here:

- ....
- ....
- ....

Keep these goals in mind each time you find yourself tempted by digital distractions.

You might even want to write them in a document on your computer or phone, so you can open them instead each time you are tempted to click on digital distractions.

COULD YOU GO SIX MONTHS?

If you need inspiration about taking a tech cleanse, I would highly recommend *The Winter of our Disconnect* by Susan Maushart.[11]

The entertaining, intelligent account charts how a mother and three teenagers (aged 14, 15 and 18) switched off for six months. Beforehand, they had what Maushart describes as a 24/7 lifestyle that had featured 'cruising for eBay bargains at midnight, posting status updates at 4 am and sleeping with phones under our pillows "just in case" (of what? falling finally into REM sleep?)'.

Maushart would wake up at 2 am to find her youngest daughter still glued to her MacBook (often still in her school uniform). Her daughter would wordlessly surrender the laptop. Maushart describes the incredible transformation her daughter, 'our family's most militant multitasker' went through once she disconnected, with dramatic changes to sleep patterns, energy levels and mood.

In a diary entry for 11 January Maushart writes:

> 11 pm. Girls talking and reading by candlelight, in clean, aired room, not glued to Facebook in zombi-ish oblivion to surrounding chaos. They are tired – as they should be at this hour – not wired.

At one point she talks about an evening spent looking at old photos ('hard copies!') devouring images and hooting and laughing.

If they'd looked at the pictures online, they might also have laughed, she said. 'But sitting side by side, passing pictures from one set of hands to another, created a different energy. We didn't simply consume the images, or allow them to consume us.'

The results of 'The Experiment'?

Six months without screens led them 'to reconnect with life itself', binding them together as a family and propelling them 'outwards and upwards as individuals'.

AIM HIGH

Ready to aim high?

Just remind yourself once again of your aims.

What is it that you want this new-found focus for?

Now take that idea, and push it. Now push it even further. What would the same idea look like if you were going for gold, not silver or bronze?

What would you aim to achieve if there really was no limit?

Write down your goal here.

- ....
- ....
- ....

For example, say your goal was to take up painting. Pushing that idea to gold could mean wanting to fill a gallery with your paintings. Or say you wanted to be happier in your job. Push it and you've got a job so perfect it could have been made for you, with inspiring colleagues and great opportunities.

Take a minute now to think about what gold would mean, and to write your answer in the space if you haven't already.

GET RUTHLESS

Have you got your golden goal?

Great.

Here's the good news: you've got enough time to achieve it.

But only if you get ruthless.

You've upped your ambition, that's the first major step.

Now you need to become ruthless about where you place your attention.

Is it a distraction? Put it to one side.

Do you really want to read about it for even a minute of your life? No? Then, it's not worth your time.

Start focusing on only the things that count to you, only the things you see as valuable information. Does it take you nearer to your goal? Yes? Then carry on...

It's your attention – no one else's. It's a valuable commodity you might not have maximised until this point.

How far can you go in avoiding digital distraction?

How much are you prepared to change?

Now take that idea, and push it – to the same degree that you pushed your goals.

If you were thinking about cutting down and checking email four times a day, cut that in half, and make it twice a day.

If you had allocated 10 minutes an hour for social media, think again. Allocate 30 minutes a day, full stop.

This might seem a harsh regime.

But sometimes you have to call into question the way things are being done by everyone else if you want to innovate, create and achieve. This is one of those times.

NINE STEPS TO FOCUS

You've just climbed to step seven. Great progress!

How do you feel now you're this far up the nine steps?

STEP SEVEN

1 *What success have you noticed recently?*

2 *Have you noticed anything else recently (about your mood, your productivity, or other people around you, for example)?*

3 *Have you tried a tech cleanse? If so, how did it go?*

A QUICK RECAP

- Read about what other people felt like when they tried a tech cleanse. The benefits – and the pain – of switching off are well documented.
- Be realistic about how difficult it will be to change your habits. Resisting the urge to check media is harder than resisting many other urges.
- Use practical tips to build up your focus stamina, such as finding a role model and setting goals.
- Push yourself and experiment with a tech detox – whether this is for 15 minutes or five days. Everyone starts from their own point.
- Think about what switching off made you feel like. Did you come up against any hurdles?
- In an inspiring account of a six-month switch off, Susan Maushart charts the highs and lows of disconnecting.

- Remember your goals. This is why you need to keep working on beating digital distractions. Expand your goals and expand your commitment to cutting back on distractions.

FURTHER READING

- Susan Maushart's book, *The Winter of our Disconnect*, is a great read that mixes a personal story with thought-provoking discussion about the 'must be constantly connected' debate. Maushart, S. (2010) *The Winter of our Disconnect*. North Sydney: Bantam.
- Just got five minutes? Read instead this article by Stuart Hughes. Got 15 minutes? Follow some of the links in the piece, discussing the surrounding debate. Hughes, S. 'Cold turkey on a "digital detox"', *BBC news magazine*, http://www.bbc.co.uk/news/magazine-19283726, 18 August 2012.

CHAPTER 11

Reboot and get balanced

Fast is busy, controlling, aggressive, hurried, analytical, stressed, superficial, impatient, active, quantity-over-quality. Slow is the opposite: calm, careful, receptive, still, intuitive, unhurried, patient, reflective, quality-over-quantity. CARL HONORÉ

WHY DO WE NEED TO reboot our minds? With so much digital pollution around, our mental 'systems' are being quickly run down.

As we gorge on information, communicate across multiple platforms, and face near constant demands for attention, we wear ourselves out.

So we need to factor in a way to restore this energy and attention. Being revitalised gives us a reserve of inspiration and focus to call on.

If we do this, then, despite the climate of digital distraction, we can think clearly and have fresh, exciting ideas. Our focus remains laser-like.

The time you most need a reboot is when your thinking is sluggish, unoriginal and forced. You may well be adamant you can't stop to recharge (you're most likely in a state of total information overload). But it's important to step away from the screen – your batteries are down to near zero.

There is no quick fix here. Pressing 'refresh' 'refresh' 'refresh' impatiently as you wait for the page to load is not going to cut it when it comes to your mind. And recharging once won't last all year.

We all need to restore our focus repeatedly. How you do that is up to you.

HOW TO REBOOT

In the last chapter we looked at doing a tech cleanse, which really helps you stay focused.

Now, we're adding inspiration, looking for that something that taps into your being and recharges your soul.

Does that sound familiar to you?

For different people, different things provide this. It's typically something that energises you, and after which you feel 'realigned'.

For me, it's dancing salsa. For a friend of mine, it's her boxing. Another friend feels re-energised after her morning run.

In each of these cases, we lose ourselves entirely in the present moment.

All the worries, niggles and concerns of the day are gone while we're dancing, boxing or running.

Our minds are so concentrated on what our bodies are doing that they switch off from everything else. We've gone out of the mind and into the body.

And though we're using up heaps of energy, at the end of it, we feel recharged. Our minds have had a real break.

Have you ever felt like this?

What gives you this feeling?

QUIET MIND

An alternative, and popular, way to refocus the mind is through yoga or meditation.[1]

As Matthew Johnstone, author of *Quiet the Mind* points out, the brain never shuts up – we can process up to 70,000 thoughts in 24 hours (that's two for every heartbeat).[2]

So, quietening the mind down is a skill we could really do with.

Johnstone says we are finding it more difficult to just sit still for 10 minutes because of the temptation to always be connected.

He points out that the Internet, smartphones, social networking and email (along with other technology) have gobbled up every waking moment with the need to be doing or saying something.

His illustrated guide to quietening the mind (see Further Reading at the end of the chapter) is a great starting point to meditation.

Don't worry if it takes a bit of time to get used to whichever form of relaxation you choose. As mindfulness and relaxation expert Shamash Alidina says:[3]

> Moving from a stressed, frantic lifestyle to one of relaxation and calm focus isn't an instant process. It has taken years to build up the various habits and tendencies that compound the stress in your life, so it will take some time to undo the stress – perhaps not years but at least a few months.

Yoga is a popular way to recharge and focus. Digital wellbeing expert, Sinead Mac Manus, has some lessons from the mat.

In the West, we often associate yoga with contorting our bodies, but yoga is much more than throwing some pretty shapes on a mat. At its heart, it is a deeply practical philosophy for living a better life and its principles and practices can help those of us living busy 24/7 connected lives to be more relaxed, focused and positive at work.

Yoga is a Sanskrit word meaning 'to yoke', and is often described as a union between body and mind, breath and movement, consciousness and impulses.

The practice of yoga works on both the body and mind in a number of ways. The most obvious one is in a physical sense – if we spend the majority of our waking hours sitting in front of a computer, then any physical movement is going to be hugely beneficial. But the benefits of yoga are also more subtle.

Yoga, and especially the deep breathing that accompanies the practice, is a powerful way of switching off the 'fight or flight' stress reflex in the body and activating the parasympathetic nervous system or 'rest and digest' system. Just three rounds of deep belly breathing can be all you need to kick your body out of stress and back into balance.

Yoga helps build our muscle of attention and our ability to focus. When our attention wanders off the breath, and off the mat, we gently bring our focus back. Over the 11 years I have been practising yoga, this skill of bringing focus back to what I am doing over and over again has meant that I am able to work in a much more focused way than many.

Yes, I still get distracted by emails, social media and other tasks, but now I am able to recognise these distractions more quickly and gently bring my focus back to my important work.

We can also learn from the rituals and routines of our yoga practice. The daily practice of getting on the mat or the meditation cushion is a great way of instilling new habits into our life and making lasting behavioural change. How can you use a series of morning, working and evening rituals to focus your day, and ingrain positive working practices?

Lastly, try adding a little Savasana to your work practice. Savasana, or corpse pose, is the lying on the floor bit at the end of class. You can literally get on the floor and just let go, even for a few minutes (perhaps one for the home office!). Or just make an intentional physical and mental break, after working intensely. You will come back to your work refreshed and refocused.

Namaste.

Sinead Mac Manus is a digital wellbeing and productivity expert, and author of *The Business Yogi: How to be Happy at Work*. Find out more at http://www.eightfold.org/ or https://twitter.com/#!/sineadmacmanus.

IN YOUR DIARY

Building a weekly refocus activity into your schedule – and also building a barbed-wire fence around that appointment in your diary – is a great way to make sure you always get a chance to recharge.

By making it a habit you are less likely to skip it.

What's going to help you recharge?

When are you going to start it?

What else might you be curious to try?

BALANCING ACT

Even if we've developed strong focus tactics, sometimes we topple off balance, and we end up losing an hour, or a day, or feeling frazzled.

Thinking about pacing can help us get balanced.

If you know you have a day full of digital noise (with an overflowing inbox and a pile of Internet research to get through) don't plan an evening of Internet shopping.

On the other hand, if you've had a quiet day of focused work, you might be ready to spend time socialising – in person or online – in the evening.

It's important to leave spaces for nothing and for silence, too.

Doing nothing helps us to recharge, and is vital if we want to stay in balance. It's harder than it sounds. You can almost guarantee that other people will try and cajole you into spending your energy on them just at the point you most need some 'nothing' time for you.

But it's often in these non-assigned moments that we notice things. Noticing dewdrops on a blade of grass, or the way light catches a treetop, or hearing a river babble can really help restore you. We'll look at the power of nature later on in the chapter.

WIRED VERSUS UNWIRED TIME

One way to pace yourself and stay balanced is to think of time as either wired or unwired.

To thrive (not just survive) in a digital world, says Tom Chatfield, author of *How to Thrive in the Digital Age*, we need to think of our lives like this:[4]

> If we are to get the most out of both the world around us and each other, we need to recognise that we now have two fundamentally different ways of being in the world: our wired and our unwired states.

Most of us experience times when we need to be connected, and communicating, and times when – now we know how to – we want to switch off.

Chatfield suggests weaving in the switched-off element every day:[5]

> Unplugged time has the most to offer us as part of our everyday living: the decision not to send emails for a morning, to turn off all phones during a meeting or meal, to set some days or hours aside for off-grid reflection, or simply to meet someone in person rather than exchanging a 20-email chain.

Write down your answers to these questions:

• What do you want the balance to be between wired and unwired time?
• When is it particularly important to be unwired?
• When is it particularly important for you to be wired?

FAST VERSUS SLOW

Another way to pace yourself is to think of things as fast versus slow.

It's easy to lose sight of the value of slow in a society, where, as go-slow expert Carl Honoré puts it: 'Reaction, rather than reflection, is the order of the day'.[6]

In his book, *In Praise of Slow*, he says that fast thinking is what we do under pressure when the clock is ticking. Slow thinking is what we do when we have time to let ideas 'simmer on the back burner'. It is intuitive, woolly and creative. 'It yields rich and subtle insights,' he says.

> True, the brain can work wonders in high gear. But it will do so much more if given the chance to slow down from time to time. Shifting the mind into lower gear can bring better health, inner calm, enhanced concentration and the ability to think more creatively. CARL HONORÉ

We seem to have created an instant response culture, where no time is taken for reflection. Slow down even a fraction before you respond, and you may find you have different ideas.

Richard Watson, author of *Future Minds*, says: 'We need to develop unhurried minds'. He points out that in the digital age deep thinking needs to be deliberately done.

He asked 999 people where they did their best thinking.[7] The top four responses were:

1 When I'm alone.

2 Last thing at night/in bed.

3 In the shower.

4 First thing in the morning.

Where do you do your best thinking?

Is it at a point when you are going fast or slow?

NATURAL BALANCE

Spending time in nature is another great way to restore your focus.

> Imagine a therapy that had no known side effects, was readily available, and could improve your cognitive functioning at zero cost. Such a therapy has been known to philosophers, writers, and laypeople alike: interacting with nature.

Researchers at the University of Michigan showed that people learn better after walking in the woods than after walking on a busy street.[8]

Their research follows on from, and backs up, Attention Restoration Theory (ART), which shows that being in nature is a proven way to restore attention.[9]

Why is this?

A different kind of attention is used when in nature, as opposed to when you are in an urban setting.

Urban environments 'are filled with stimulation that captures attention dramatically and additionally requires directed attention (e.g. to avoid being hit by a car), making them less restorative'.[10]

In nature, this 'directed attention' gets a chance to be replenished, as it is not being used. The type of attention used instead is what is termed 'involuntary attention', which is captured by intriguing stimuli like a sunset.

The researchers found that, even just looking at pictures of nature (instead of pictures of the city), meant people performed better in directed-attention tasks, as their attention had been restored.

One of the things directed attention is used for is suppressing distracting stimulation – something we have been dealing with a lot! So, knowing how to restore this attention is vital.

There are simple ways you can use this in digital life. For example, as an antidote to the digital and the urban, I have a screensaver of an image I took while immersed in nature.

Could you set up as a screensaver a natural image you have taken?

BIG CITY LIVING

In many ways, our digital lives are like living in a big city.

The way our attention is pulled around in the digital world is similar to the way it is pulled around in an urban environment. We must work to ignore distractions, our attention is being captured in an aggressive way, and the outlook is not calming.

The world we live in digitally is certainly urban, not natural. So we must be aware that it depletes rather than restores our attention.

Comparisons to big-city living can be drawn in a social sense, too. People are so close to one another yet at the same time so distant (like on a crowded tube in London just as on social media). We may see an awful lot of people yet speak to no one. We may feel like we don't have enough space to ourselves at the same time as feeling alone.

We keep our distance by avoiding eye contact or putting on our headphones in the real city. We text instead of talk, email instead of write, Facebook a friend instead of visit them. We are marking our boundaries, which, in some ways, we are pulling in ever closer to us.

But being so close and yet so far – physically or virtually – can make us feel more alone than if we were walking for miles in the country not meeting a soul.

We need to remember what the big city of the digital world is doing to our attention, and make sure we counteract that by doing things that restore us.

NINE STEPS TO FOCUS

Almost there! You've made it to step eight. You've done brilliantly to climb this far.

How are you feeling about your progress?

STEP EIGHT

1. *What progress do you feel you've made since you began this journey?*
2. *How has your digital behaviour changed?*
3. *What have the changes you have made meant for other areas of your life?*

A QUICK RECAP

- How could you refresh your energy and attention?
- What restores you? How often do you make time for this?

- What are your feelings about yoga or meditation? Have you ever tried them as a way to refocus?
- Thinking about wired and unwired time is one way to stay balanced.
- As a contrast to the fast-paced digital world, slow down. This new pace could bring fresh ideas.
- Spending time in nature helps refresh your mind, according to Attention Restoration Theory. So get out there!
- Does the digital world ever feel like big-city living to you?

FURTHER READING

- For a short, easy introduction to meditation I recommend Matthew Johnstone's beautifully illustrated *Quiet the Mind*. Johnstone, M. (2012) *Quiet the Mind*. London: Constable & Robinson Ltd.
- *How to Thrive in the Digital Age* is a little book you can pretty much fit in your pocket. A good alternative to checking a screen in those spare moments. Chatfield, T. (2012) *How to Thrive in the Digital Age*. London: Pan Macmillan.

CHAPTER 12

The ups, the downs, the future

Unplugging each Friday night with my family is now the day that I rush towards each week. The 24 hours not online truly resets my soul. TIFFANY SCHLAIN

THE UPS

Seriously – the quality of the silence has changed. It's thicker, more meditative. The buzz is gone. It's good. SUSAN MAUSHART[1]

HOW HEALTHY IT IS TO DISCONNECT, and focus. How welcome is what was already there, that we had lost sight of.

- You'll find peace.
- You'll sleep better.
- You'll be more creative.
- You'll have better ideas.
- You'll be less stressed.

- You'll listen properly to those around you.
- You'll make and recall more memories.
- You'll think more deeply.
- You'll enjoy being absorbed in a task.
- You'll improve your relationships.
- Life will seem a lot easier.

Not a bad payoff for simply redirecting your attention.

It's time to reintroduce technology on your terms, and enjoy using it.

Think about the following questions. Writing down your answers will help you process the ideas.

Notice any areas where your thinking has changed since you began working on tackling digital distraction.

1	Which digital areas are most important to you? Make a list (e.g. social media platforms, specific websites, certain news feeds or blogs.)
2	Are there any digital platforms that you use that you dislike? Can you quit these if you haven't already?
3	What time of day do you most often use digital platforms? Does this time of day work well for you?

4 Do you have rules or policies for controlling digital platforms? List these. If you don't have any rules yet, can you make some?
(Examples of rules might be 45 minutes' work/15 minutes' distraction, 55 minutes/five minutes, email three times a day, mobile off after 8 pm, etc.)

Having a policy for digital use shifts the control. You consume digital media rather than being consumed by its distractions.

We want to be able to take advantage of the opportunity technology offers us.

But not at the expense of our memories, thoughts and relationships.

> Without the ability to say 'no' as well as 'yes' to technology we risk turning its miracles into snares. TOM CHATFIELD[2]

THE DOWNS

The downside most people fear when contemplating switching off is missing out.

This fear stops us switching off our phones, it keeps us glued to social media, and it means we let emails ping at us all the time. We worry we might miss knowing what is going on, lose out socially, or miss out on information we need.

Did you have this fear initially?
How did it come about?

Most people realise, when they disconnect, that they are not missing out and that what was previously deemed urgent, isn't.

But while missing out doesn't tend to cause us the grief we had thought it would, other things do cause difficulties.

One of the hardest things about switching off initially is coping with boredom. We have become used to reaching for digital distractions to fill any space we are dealt, any gap in our time.

How should we fill all that time?

If we leave gaps, we are forced to invent ways to fill them, and left to come up with something to do. This can be uncomfortable, and we can feel bored.

Did you experience this?

If you work through the boredom then, often, thoughts and ideas start to arrive.

> Boredom is beautiful. Rumination is a prelude to creation. Not only is doing nothing one of life's few remaining luxuries, it is also a state of mind that allows us to let go of the external world and explore what's deep inside our head. But you can't do this if 10 people keep sending you messages about what they are eating for lunch... RICHARD WATSON[3]

Being focused all the time can also be very tiring – in an entirely different way from the drained tiredness you get from being constantly distracted.

Another difficult feeling we may experience is irritability when we go back to using digital media. Our tolerance for long-winded emails, pointless status updates and boring Internet pages will have gone right down – we don't want people wasting our time now. We hand out our attention purposefully and carefully, rather than in the reckless way we used to spend it.

There are other downsides you may have experienced, such as feeling out of sync with others, or being annoyed at having to log back in to do a certain task that requires email.

What downsides did you experience when you disconnected?

1
2
3

Were these different to the negative consequences you expected?

REWIND

When things get hard, rewind for a moment.

Ask yourself, where were you at the start of this book?

Were you hunched over your computer or smartphone at all hours of day and night, eyes glazed?

Were you talking at the same time as posting status updates, writing a report, and downloading music, never giving your full attention to any one thing?

How does this compare to where you are now?

Sometimes when we step ahead onto new ground, the walking there becomes easy. We forget that it was ever difficult to take that first step.

Remember for a moment how far you've come, and celebrate it.

PRESS PAUSE

Procrastination is nothing new, it's just that now we've got a new range of distracting tasks. Cleaning the kitchen is posting on social media, tidying out the cupboards is randomly surfing the web.

We're good at tricking ourselves not to get on with things. Nothing changed just because the Internet came along.

Can these interruptions ever be welcome? The good news is yes.

Procrastination often happens when we start a task or restart after a break.

Notice what you are doing. Laugh at yourself, almost, as you see yourself building up to finally starting on what you were meant to be doing. So you think…

Oh, it's suddenly urgent to search for the average temperatures in Greece in July is it, in November… and now I want to research the price of a new kitchen, even though I have no intention of buying one… procrastination… I give myself three more random Internet searches and then I'll begin.

Stick to your word and, after three more random Internet searches, you might find you're ready to do your task. You've built up to it.

Digital distraction also comes, however, when you're mid-task. Suddenly you have an urge to check email, or browse the net.

As long as you're adept at understanding and recognising your own distraction (which you hopefully will be by now), sometimes, a short interruption is useful.

The first step to take is to acknowledge that you are distracted.

The next step is to decide with yourself how long you will stay distracted. Five minutes is fine. That's a bit like (in non-digital terms) going to make a cup of tea – a welcome, yet short, interruption.

If you don't keep the interruption short, there is a danger of it spiralling into a lengthy journey of distraction. We know only too well how one link can lead to another and 'just sending a quick email' can be an hour wading around in your inbox.

A useful interruption is more like a quick trip round the block than taking the turning for the motorway and easing into fifth.

If you end up miles from where you began, it's a long trek back to get started again. So if you do want to let in the odd distraction, watch the clock.

FAST FORWARD

What does the digital world we live in look like?

If you were to paint it, or draw it, how would it look to you?

Would the landscape be underground? Would it smell of earth? Would there be colour or greys and blacks? What would it sound like? Feel like?

How would you represent the Internet, mobiles, social media, or emails if you drew them?

> *Take some time to think about the landscape you see, or sketch it out now.*

Now, fast forward. What has happened to your landscape in five years' time? What does it look like in 10 years' time?

The pace of digital change that we have lived through in the past two decades has been immense. We've switched to digital lives, completely changed our behaviour, and can hardly remember how things used to be.

But before platforms such as email and social media were invented, we couldn't imagine what they would look like.

So, what happens next?

Experts are predicting turbulent times as we learn to cope with the media landscape we have created.

Joi Ito, the director of MIT Media Lab, uses an immune system metaphor. He points out that, at this point, we are just infants. He says we are going to see a lot of sickness and a lot of bumps but that we will survive and technology will stay. The biggest flu coming up is the privacy problem, he says.[4]

Writer Maggie Jackson sees the future as bleak, and warns against eroding our capacity for deep, sustained, perceptive attention. 'We are on the verge of losing our capacity as a society for deep, sustained focus,' she cautions.[5]

Author Richard Watson makes 10 predictions for the future, including that:

> Digital storage will allow us to record our entire lives via wearable devices. Our lives will therefore become fully searchable. This will have a range of implications, ranging from 'memory theft' through to issues relating to the death of forgetting.

He also says that machines will become aware of the emotional state of their users and humans will become more machine-like.[6]

Another vision of the future is given by Virgin Media, which has created a video showing what digital life could be like in 2025. People conjure up screens in the bathroom mirror, in fields, and at kitchen tables, and are told by virtual experts how to cook, how much water they should drink, and what medication to take.[7]

I think people are about to become much more discerning about where they place their attention. We will become less tolerant of companies that prey on us – wanting our attention – and of individuals who waste our time. The battle for our attention will intensify, and we will be willing to pay for spam-free zones.

Digital-free destinations and digital-free time will become a normal concept. However, before that, total information overload will continue to cause much distress.

New communication systems will develop to accommodate an increased value placed on attention and focus.

And, after going further down the line of distancing ourselves from each other, we will see a resurgence of face-to-face communication, and a higher value placed on this.

For the generation who have grown up knowing only a connected world, there will be most change. They have been early adopters, but they are also likely to be a big part of a backlash against constant connectivity – they are the demographic most hit by all of this.

Their ways of thinking will determine what happens to society in future, and so we must invest in them. They can teach us about the future, but we must also teach them about the past, before vital skills such as deep thinking are forgotten.

Tony Hughes sees hope for the future of technology – but only if we entrust it to the young.

Before the Internet it was television. And, if not that, it was radio, films, or games. All have taken their turn as the popular bogeyman, blighting the minds of the young.

But the Internet brings with it an additional threat – children can access it 24/7, through increasingly smarter phones, connected televisions, and tablet PCs.

One accusation is that the Internet is detrimental to developing minds – that should be rote learning from books, and listening rapt to teachers.

Technology has become a byword for all that is wrong with the young and their apparent lack of education.

But we don't hear alarmist cries when new technology is introduced into hospital surgeries, cars, or railways. We frame those advances in terms of progress. So why the discrepancy when new technology is used by children in our homes, and introduced into the classroom?

The answer often given is: 'Well, of course, it depends on how it is used'. This is exactly my point. It always depends on how it is used. Technology itself is neither good nor bad – we can use it to maim or mend.

The Internet has democratised information – the key to power – and placed the largest database of human knowledge at our fingertips.

Last year, a 15-year-old American teenager, whilst bored during a biology class, came up with an idea to improve cancer tests. Using the Internet, he scoured scientific journals, and produced a method for testing for a number of common cancers that has proved many times faster, and cheaper, than anything to date.

He contacted, using email, 200 academics renowned in the field. After 199 rejections, he had one positive response to help him develop his technology.[8]

This is an amazing and uncommon use of new technology, but it demonstrates how children can, and have, used the most sophisticated means of human communication on this planet to improve the lives of us all.

Children are rightly distracted by new technologies, as they intuitively understand the potential there is. They are open to possibilities in a way that often contrasts and clashes with their parents and teachers.

The views of their parents and teachers are often formed out of fear of the new, rather than a reasoned critique informed by knowledge of the technology.

Teachers perceive the introduction of technology to the classroom as a move to replace them, whilst parents fear what their sons and daughters do when unsupervised.

Both blame technology for the poor behaviour of children, yet often they are not willing to engage with the technology themselves.

The problem with new technologies is not the technology itself, but the approach we often take, defining technology as detrimental to the very youth that understands best how it can improve lives and not destroy them.

Technology continues to improve, yet our age-old response to new advances remains the same.

Tony Hughes is a Director of Focus Innovation and One of Us, working with traditional industries to introduce new technologies. See www.focusinnovation.co.uk and www.oneofus.me for more information.

THE SOUNDTRACK TO YOUR FUTURE

Beeping, buzzing and pinging are part of our sound landscape now – in every house and on every street, just another bit of the music of daily life.

The noise will keep coming: this is the soundtrack to our future.

But, if you already know how to turn down the volume on digital distractions, you'll be better equipped as things get even louder.

What will your digital future look like? What would you like it to hold?

Think about the work you have done during this book:

What worked easily for you?

What was really, really hard?

Is there one type of digital distraction that you still can't beat? Have you changed your views about different types of digital platforms? Have you changed your habits? How?

Take some time to think back now about your progress, and note down things that you have noticed along the way.

Also take some time to make yourself a plan – think about what next steps you want to take (for example, you might want to disconnect one day a week/do a digital detox holiday/have a digital clearout).

Progress noticed:
1

2
3
4
5
Next steps:
1
2
3

Now, break down those steps and decide when you want to achieve them by. Set yourself a target for three months' time, six months' time, and a year from now.

In three months .... In six months .... In one year ....

NINE STEPS TO FOCUS

Well done! Congratulations! Time to celebrate!

You've reached step nine, and the end of the practical stage of the course tackling digital distraction. Brilliant work! Give yourself time to look back, time to appreciate how far you've come, and time to celebrate.

STEP NINE

1 *Can you remember what your digital habits were like at the start of this course? (Look back at your notes to remind yourself.) Celebrate how far you've come.*

2 *How do you feel now, as the course comes to an end?*

3 *If you were to really go for it, what one other thing would you try?*

A QUICK RECAP

- Sleeping better, being more creative and feeling healthier are just some of the major benefits of being focused.
- Fear of missing out is one common reason for not wanting to switch off. In actuality, boredom and tiredness might be bigger hurdles you encountered.
- Don't forget how far you've come on the days when beating distractions gets tough.

- We've always been good at procrastination. But digital devices give us a new way to put things off. Sometimes we can use this procrastination positively.
- Have your views changed about any digital platforms?
- Limit your distractions to the equivalent of a walk round the block, so you don't end up hundreds of miles down the motorway.
- What does the future hold? Experts are predicting stormy times ahead as we get used to our digital landscape.
- What does your future hold? Make some plans for your next steps tackling digital distraction.

FURTHER READING

- Turn straight to the back of Richard Watson's *Future Minds* to read his 10 predictions for the digital future. Watson, R. (2010) *Future Minds: How the Digital Age is Changing Our Minds, Why This Matters and What We Can Do About It*. London, Boston: Nicholas Brealey Publishing.
- Want a visual representation of the future? This video from Virgin Media predicts that things will look very different relatively soon: http://www.virginmediabusiness.co.uk/generationip/.

Tuning in to a digital life

What did you trade the minutes of your life for? Hopefully it was happiness. MARK LACAS

LET'S GO OLD-SCHOOL DIGITAL DISTRACTION for a minute – as you switch between radio stations trying to find something you want to listen to…

'For one weekend only in our half-price sale…' 'And with the latest headlines…' 'A bright start, with showers later and highs of 13 degrees…'

Your brain immediately starts working hard processing the information.

Fast-forward to today, and you're doing this kind of 'listening' across digital platforms on a near constant basis: 'Tweet tweet' 'cc-ing you in…' 'Beep beep…'

You switch between frequencies with snippets of information constantly streaming at you. There comes a point when your attention wanes, and all you hear is noise.

If we don't determine where we want to spend our attention, other people will quickly decide for us. An information stream will already be flowing, just waiting to direct our thoughts and take our time if we happen to tune in.

But if we say 'not now', rather than 'always on', we get, instead, time to think for ourselves. If we know how to tune in and out of digital life, our own lives stay balanced.

HOW THE MAGIC STARTS

Then, when we focus our attention, the magic starts.

Giving 100 per cent of your attention to something, or someone, brings deeper connections and 'ah ha' moments.

Particularly in a culture where multitasking has become the norm, you set yourself apart in this simple way – just by paying attention.

Spreading your attention thinly means no one gets the best of you. Life is littered with unfinished business and mistakes.

Instead, giving your full attention to things means task after task gets completed with seeming ease. You also improve your relationships and collect vivid memories as you go.

On a personal level, being present, and focused on each thing, is a much more satisfying way to live. It's less stressful and you feel in control.

On a professional level, being focused means you reach goals, go further than you thought you could, and have time to take big steps.

On a wider economic scale, productivity improves and money is made. Too often, in the age of distraction, money is wasted paying workers for unproductive hour after hour.

IS THIS THE BEGINNING OR THE END?

As you start to pass on what you've learnt about digital distraction, it's time for wider questions, and to think about the impact on society.

What do we most need technology for? How do we want to use it?

You might find yourself discussing these questions with others.

What tasks can technology do for us? What do we want to keep doing for ourselves? How best can we communicate with each other through technology?

We didn't ask these questions at the start of that whirl when we invited email, smartphones, the Internet and social media into our homes.

We couldn't ask those questions then because we weren't familiar with the digital platforms. We didn't know how to use them, never mind what we needed them for.

But now that they are part of what knits us together as a society, it's our responsibility to question what we want from them.

Are we using technology in the right way? Is it controlling us or are we controlling it? Have we adapted to it or has it adapted to our needs?

A wider debate is just beginning. I hope you will be part of it.

A ROAD WELL-TRAVELLED

As we come to the end of this particular road, it's time to take one last look around.

By now, you are familiar with the landscape we are

travelling through. You have been journeying this way long enough to know the landmarks.

Nod as you see them – email, social media, the Internet, and smartphones – and expect that they will try to pull you off track.

Sometimes they still manage it, yet you know that distraction cycle well, and you know it won't last.

The more familiar the focused route begins to feel under your feet, and in your mind, the easier it becomes to travel.

So, in the coming weeks, months, and years, I hope you continue to determine your own digital consumption, rather than being consumed by distraction.

I also hope that you enjoy the thinking time you've won, to dream and make exciting plans.

Here it is then, the end of the road. I hope you've had a great journey.

For me, it's been an absolute pleasure.

Each time I sat down to write, I switched off all digital distractions. By the end, I decided to switch off entirely for three days (taking my cue from the Utah neuroscientists).

I put my phone away in a box (something I've never done before) and, for 72 hours, I wrote, I thought, and I finished. Time seemed long and plentiful, with no tugs at my consciousness.

I'm about to log back in now to the 'beep beep' 'pling' 'buzz' digital orchestra, and I'm steeling myself for it.

How about you? Are you ready to switch back on?

Before you do, stop for a second and look forward to the next time you'll deliberately switch off.

Why?

Because you know how to conduct the orchestra now, rather than covering your ears and wincing at the noise.

References

PART 1

Chapter 1

Opening quote: Risner, N. (2006) *The Impact Code: Live the Life You Deserve*. John Wiley & Sons.

1 A short history of email timeline, http://www.macworld.com/article/1167303/timeline_a_brief_history_of_email.html.

2 Research in motion study shows that in November 2004 there were more than two million BlackBerry subscribers worldwide. There were three million by May 2005: http://www.bbgeeks.com/blackberry-guides/the-history-of-the-blackberry-88296/.

3 Price, I. 'Three cheers for four-day weeks', *The Guardian*, 16 April 2011.

4 Carr, N. (2010) *The Shallows: How the Internet is Changing the Way We Read, Think and Remember*. London: Atlantic Books.

5 Cited in Carr, N. (2010) *The Shallows: How the Internet is Changing the Way We Read, Think and Remember.* London: Atlantic Books

6 *Tonight*, ITV. Study discussed by Susan Greenfield, 'Is technology taking over our lives?', *Tonight*, ITV, 18 October 2012.

7 Carr, N. (2010) *The Shallows: How the Internet is Changing the Way We Read, Think and Remember*. London: Atlantic Books.

8 An international study led by the University of Cambridge found that 38 per cent of 10- to 18-year-olds feel overwhelmed by communications technology (34 per cent for 25 to 34-year-olds). 36 per cent of adults and 43 per cent of young people are taking steps to limit their usage.

9 Study by Microsoft Corporation cited in Bregman, P. (2011) *18 Minutes: Find Your Focus, Master Distraction, and Get the Right Things Done*. London: Orion.

10 Rosen, L. D. (2012) *iDisorder: Understanding Our Obsession With Technology and Overcoming Its Hold On Us*. New York: Palgrave Macmillan.

11 Rideout, V. J., Foehr, U. G. and Roberts, D. F. 'Generation M2: Media in the lives of 8- to 18-year-olds', A Kaiser Family Foundation Study, January 2010, http://www.kff.org/entmedia/8010.cfm.

12 Sawyer, M. 'State of play', *The Observer Magazine*, 15 July 2012.

13 Shlain, T. 'National day of unplugging: a digital detox', *Huffington Post*, 23 March 2012: 'She watched her 2-year-old scroll a block of wood like it was an iPad and says maybe that was when she realised she needed to unplug.'

14 Rosen, L. D. (2010) *Rewired: Understanding the iGeneration and the Way They Learn*. New York: Palgrave Macmillan.

15 Turkle, S. (2011) *Alone Together: Why We Expect More From Technology and Less From Each Other*. New York: Basic Books.

16 Ophir, E., Nass, C. and Wagner, A. D. 'Cognitive control in media multitaskers', *Proceedings of the National Academy of Sciences*, published online before print, 24 August 2009.

17 Video interview with professor David Strayer at http://www.pbs.org/newshour/extra/video/blog/2009/07/texting_and_driving_a_dangerou.html. Data he discusses comes from crash data and driving simulation experiments.

18 1,152 people were treated in 2011 in the USA, according to Consumer Product Safety Commission, from http://www.huffingtonpost.com/2012/07/30/texting-while-walking_n_1717864.html.

19 'Is technology taking over our lives?', *Tonight*, ITV, 18 October 2012.

20 Ibid.

Chapter 2

Opening quote: Turkle, S. (2011) *Alone Together: Why We Expect More From Technology and Less From Each Other*. New York: Basic Books.

1 Jakob Nielsen, http://www.useit.com/.

2 Carr, N. (2010) *The Shallows: How the Internet is Changing The Way We Read, Think and Remember*. London: Atlantic Books.

3 Young, K. (1998) *Caught in the Net: How to Recognise the Signs of Internet Addiction – and a Winning Strategy for Recovery*. New York: Wiley.

4 Ofcom's 2011 Communications Market Report.

5 Godson, S. 'Digital romance', *Psychologies Magazine*, July 2012.

6 Kirwan-Taylor, H. (2012) 'Are you sure you want to shut down?', *ES Magazine*.

7 Carr, N. (2010) *The Shallows: How the Internet is Changing the Way We Read, Think and Remember*. London: Atlantic Books.

8 Borg, J. (2010) *Mind Power: Change Your Thinking, Change Your Life*. Harlow: Pearson Education Limited.

9 Sparrow, B., Liu, J. and Wegner, D. 'Google Effects on Memory: Cognitive Consequences of Having Information at Our Fingertips', *Science*, 5 August 2011, Vol. 333, No. 6043, pp. 776–8 and 'The Digital Human', Radio 4, 15 October 2012.

10 Ofcom's 2011 Communications Market Report shows 81 per cent of smartphone users have their phones switched on all the time and 38 per cent of adult users and 40 per cent of teen users admit to using their smartphone after it woke them.

11 Hodgekiss, A. 'The real reason you're not sleeping', *Psychologies Magazine*, June 2012.

12 Babauta, L. (2010) *Focus: A Simplicity Manifesto in the Age of Distraction*, ebook.

13 Maushart, S. (2010) *The Winter of Our Disconnect*. North Sydney: Bantam.

14 Greenfield, S. (2008) ID: *The Quest for Identity in the 21st Century*. Great Britain: Sceptre.

15 Jackson, M. (2008) *Distracted: The Erosion of Attention and the Coming of the Dark Age*. New York: Prometheus Books.

16 Turkle, S. (2011) *Alone Together: Why We Expect More From Technology and Less From Each Other*. New York: Basic Books.

17 Ofcom's 2011 Communications Market Report shows that 58 per cent of people use text messages to communicate with family and friends at least once a day, compared to 49 per cent communicating face-to-face.

18 Abel, J. 'The rise of digital detox', *Forbes*, 15 May 2012, http://www.forbes.com/sites/gyro/2012/05/15/the-rise-of-digital-detox/.

19 For users with 120 friends, the average man interacts with four people and the average woman with six, so I have used five as an overall average. 'Primates on Facebook', *The Economist*, 26 February 2009.

Chapter 3

Opening quote: Babauta, L. (2010) *Focus: A Simplicity Manifesto in the Age of Distraction*, ebook.

1 Baddiel, D. *Psychologies Magazine*, June 2012.

2 Nielsen, J. http://www.useit.com/.

3 Kraus, J. 'We're creating a culture of distraction', 25 May 2012, http://joekraus.com/were-creating-a-culture-of-distraction.

4 Kirwan-Taylor, H. (2012) 'Are you sure you want to shut down?', *ES Magazine*.

5 'Generation IP 2025' by Virgin Media, and a study by Virgin Media and The Future Laboratory article at http://www.virginmediabusiness.co.uk.

6 Wolverson, R. 'Game changers', *Time*, 29 March 2012.

7 'Atos boss Thierry Breton defends his internal email ban', BBC News, technology section, 6 December 2011, http://www.bbc.co.uk/news/technology-16055310.

8 Wolff, J. (2010) *Focus: Use the Power of Targeted Thinking to Get More Done*. Harlow: Pearson Education Limited (first edn 2008).

9 Email bankruptcy was a term coined by Sherry Turkle. High-profile examples include Jeff Nolan: http://jeffnolan.com/wp/2007/04/24/e-mail-bankruptcy/. Moby also famously said he was not checking email for the rest of the year.

10 Naughton, J. 'Want to improve your efficiency? Try switching off', *The New Review*, *The Observer*, 20 May 2012.

11 Babauta, L. (2009) *The Power of Less: The 6 Essential Productivity Principles That Will Change Your Life*. London: Hay House.

12 Babauta, L. (2010) *Focus: A Simplicity Manifesto in the Age of Distraction*, ebook.

13 See http://www.merlinmann.com/ and http://inboxzero.com/articles/ for more information about Inbox Zero.

14 Statistics from Ofcom's 2011 Communications Market Report.

15 In the USA, before getting out of bed in the morning, 35 per cent of people open smartphone apps, a quarter check email and 18 per cent check Facebook. Ericsson ConsumerLab (2011) data cited by Rosen, L. D. (2012) in *iDisorder: Understanding our Obsession With Technology and Overcoming Its Hold On Us*. New York: Palgrave Macmillan.

16 Price, I. 'Four-day working week? Three cheers!', *The Guardian*, 16 April 2011.

17 Statistics from Ofcom's 2011 Communications Market Report.

18 Kraus, J. 'We're creating a culture of distraction', 25 May 2012, http://joekraus.com/were-creating-a-culture-of-distraction.

19 Rosen, L. D. (2012) *iDisorder: Understanding Our Obsession With Technology and Overcoming Its Hold On Us*. New York: Palgrave Macmillan.

20 Ibid.

21 Taylor, C. 'For Millennials, Social Media Is Not All Fun and Games', 29 April 2011, GigaOM, http://gigaom.com/2011/04/29/millennial-mtv-study/.

22 Cited by Kirwan-Taylor, H. 'Are you sure you want to shut down?', *ES Magazine*, 16 March 2012.

23 Turkle, S. (2011) *Alone Together: Why We Expect More From Technology and Less From Each Other*. New York: Basic Books.

PART 2

Chapter 4

Opening quote: Greenfield, S. (2008) *ID: The Quest for Identity in the 21st Century*. Great Britain: Sceptre.

1 Aboujaoude, E. (2011) *Virtually You: The Dangerous Powers of the E-Personality*. New York, London: W. W. Norton & Company.

2 Borg, J. (2010) *Mind Power: Change Your Thinking, Change Your Life*. Harlow: Pearson Education Limited.

3 Aboujaoude, E. (2011) *Virtually You: The Dangerous Powers of the E-Personality*. New York, London: W. W. Norton & Company.

4 'S Korea child starves as parents raise virtual baby'. BBC News, 5 March 2012, http://news.bbc.co.uk/1/hi/world/asia-pacific/8551122.stm and

'Korean couple let baby starve to death while caring for virtual child', *The Daily Telegraph*, 5 March 2010, http://www.telegraph.co.uk/news/worldnews/asia/southkorea/7376178/Korean-couple-let-baby-starve-to-death-while-caring-for-virtual-child.html.

5 Fackler, M. 'In Korea, a boot camp cure for web obsession', *The New York Times*, 18 November 2007, www.nytimes.com/2007/11/18/technology/18rehab.html.

6 www.netaddiction.com.

7 Young, K. (1998) *Caught in the Net: How to Recognise the Signs of Internet Addiction – and a Winning Strategy for Recovery*. New York: Wiley.

Chapter 5

Opening quote: Davies, W. H. From the poem, *Leisure*. See the full poem at http://www.winningwordspoetry.com/poems/leisure/.

1 Kraus, J. 'We're creating a culture of distraction', http://joekraus.com/were-creating-a-culture-of-distraction, 25 May 2012.

2 Babauta, L. (2010) *Focus: A Simplicity Manifesto in the Age of Distraction*, ebook.

3 Csikszentmihalyi, M. (1991) *Flow: The Psychology of Optimal Experience*. HarperPerennial.

4 Jouvent, R. 'We are all programmed to look to the future', *Psychologies Magazine*, July 2012.

5 In a talk given by Sinead Mac Manus at Wellbeing in the City (11 August 2011), she asked everyone to mark on a chart when their most productive hours of the day were.

Chapter 6

Opening quote: Thoreau, H. D. (1962) *Walden, or Life in the Woods: On the Duty of Civil Disobedience*. New York: Macmillan Publishing Company, a division of Macmillan, Inc.

1 Though boredom is. 'Boredom – the word itself hardly existed 150 years ago – is a modern invention,' says Carl Honoré (2004) in *In Praise of Slow*. London: Orion Books.

2 Thoreau, D. H. (1962) *Walden, or Life in the Woods: On the Duty of Civil*

Disobedience. New York: Macmillan Publishing Company, a division of Macmillan, Inc.

3 Quotes and account of trip reported by Richtel, M. 'Outdoors and out of reach, studying the brain', *The New York Times*, 15 August 2012.

4 Perlow, L. (2012) *Sleeping With Your Smartphone: How to Break the 24/7 habit and Change the Way You Work*. Harvard Business Review Press.

5 'Atos boss Thierry Breton defends his internal email ban', BBC News, technology section: http://www.bbc.co.uk/news/technology-16055310, 6 December 2011.

6 Amir Khan giving advice to Team GB boxers during BBC London 2012 Olympics coverage, 10 August 2012.

7 *'During the Games period we'll be focusing on delivering our best possible performance, and trying to avoid distractions such as social media.'* Women's Match Racing team quote on www.matchracegirls.com.

8 Bagchi, R. '50 stunning Olympic moments No. 36: Linford Christie wins gold in 1992', http://www.guardian.co.uk/sport/blog/2012/jun/05/linford-christie-barcelona-olympics-1992, 5 June 2012.

9 Hartley, S. 'Athletic focus and sport psychology: key to peak performance', *Podium Sports Journal*, 9 December 2012. See http://www.podiumsportsjournal.com/2010/12/09/athletic-focus-sport-psychology-key-to-peak-performance/. Hartley also cites Taylor, J. 'Understanding focus in sports', *Psychology Today*, with reference to the torch beam idea, 13 July 2010.

Chapter 7

Opening quote: Gleick, J. (1999) *Faster*. New York: Pantheon Books.

1 Parkinson's Law as explained in http://www.economist.com/node/14116121.

2 Babauta, L. (2009) *The Power of Less: The 6 Essential Productivity Principles That Will Change Your Life*. London: Hay House.

3 Watson, R. (2010) *Future Minds: How the Digital Age is Changing Our Minds, Why This Matters and What We Can Do About It*. London, Boston: Nicholas Brealey Publishing.

4 Tracy, B. (2004) *Eat That Frog! Get More of the Important Things Done – Today!* Mobius.

5 Created by Cali Ressler and Jody Thompson. See http://www.gorowe.com/. A blogpost on presenteeism: Ressler, C. (6 February 2012) 'The costs and causes of presenteeism' is at http://www.gorowe.com/blog/2012/02/06/accountability/the-costs-and-causes-of-presenteeism/.

6 Allen, D. (2001) *Getting Things Done*. London: Piatkus Books Ltd.

7 Gleick, J. (1999) *Faster*. New York: Pantheon Books.

8 Shirky, C. 'It's Not Information Overload. It's Filter Failure', talk at Web 2.0 Expo NY, http://www.youtube.com/watch?v=LabqeJEOQyI.

9 Babauta, L. (2009) *The Power of Less: The 6 Essential Productivity Principles That Will Change Your Life*. London: Hay House.

10 For more suggestions of tools see Chapter 7 of this free ebook: Babauta, L. (2010) *Focus: A Simplicity Manifesto in the Age of Distraction*, ebook.

Chapter 8

Opening quote: Jackson, M. (2008) *Distracted: The Erosion of Attention and the Coming of the Dark Age*. New York: Prometheus Books.

1 Kirwan-Taylor, H. 'Are you sure you want to shut down?', *ES Magazine*, 16 March 2012; Thompson, D. 'Would you rather spend time with a smartphone?', *The Daily Telegraph*, 16 June 2012; Kraus, J. 'We're creating a culture of distraction', 25 May 2012, http://joekraus.com/were-creating-a-culture-of-distraction.

2 Bregman, P. (2011) *18 Minutes: Find your Focus, Master Distraction, and Get the Right Things Done*. London: Orion.

3 Kirwan-Taylor, H. 'Are you sure you want to shut down?', *ES Magazine*, 16 March 2012.

4 Jackson, M. (2008) *Distracted: The Erosion of Attention and the Coming of the Dark Age*. New York: Prometheus Books.

5 As mentioned in Part 1, some 40 per cent of the time we never get back to what we were originally doing. The more difficult the original task, the less likely we are to resume it. Study by Microsoft Corporation cited in Bregman, P. (2011) *18 Minutes: Find Your Focus, Master Distraction, and Get the Right Things Done*. London: Orion.

6 A year-long study cited in Jackson, M. (2008) *Distracted: The Erosion of Attention and the Coming of the Dark Age*. New York: Prometheus Books.

7 Bregman, P. (2011) *18 Minutes: Find Your Focus, Master Distraction, and Get the Right Things Done*. London: Orion.

8 Jackson, M. (2008) *Distracted: The Erosion of Attention and the Coming of the Dark Age*. New York: Prometheus Books.

9 Borg, J. (2010) *Mind Power: Change Your Thinking, Change Your Life*. Harlow: Pearson Education Limited.

10 Thompson, D. 'Would you rather spend time with a smartphone?', *The Daily Telegraph*, 16 June 2012.

11 Przybylski, A. K. and Weinstein, N. 'Can you connect with me now? How the presence of mobile communication technology influences face-to-face conversation quality', *Journal of Social and Personal Relationships*, online first, http://spr.sagepub.com/content/early/2012/07/17/0265407512453827.full.pdf+html, 19 July 2012.

12 As shown by Cameron Marlow's findings about interactions on Facebook. 'Primates on Facebook', *The Economist*, 26 February 2009.

13 Rosen, L. D. (2012) *iDisorder: Understanding Our Obsession With Technology and Overcoming Its Hold On Us*. New York: Palgrave Macmillan.

14 Such as in the case of Ben Goldsmith and Kate Rothschild, who played out their arguments over ending a nine-year marriage on Twitter before stating that, 'Things have been said in public which should have been kept private' and taking their Tweets down, documented in Clark, A. 'How Twitter and social media are putting an end to our private lives', *The Observer*, 10 June 2012.

Chapter 9

Opening quote: Ferriss, T. (2007) *The 4-Hour Workweek*. USA: Crown Publishing Group. (This edition 2008, Vermilion.)

1 Turkle, S. (2011) *Alone Together: Why We Expect More From Technology and Less From Each Other*. New York: Basic Books.

2 Ofcom's 2011 Communications Market Report shows that text messages are the most used method for daily communication with family and friends (58 per cent of UK adults text friends and family at least once per day compared to 47 per cent who talk on a mobile. For 16 to 24-year-olds, 90 per cent text at least once per day, and 67 per cent talk on a mobile).

3 Turkle, S. (2011) *Alone Together: Why We Expect More From Technology and Less From Each Other*. New York: Basic Books.

4 Bernstein, E. 'Your Blackberry or your wife', *The Wall Street Journal*, 11 January 2011.

5 Maushart, S. (2010) *The Winter of our Disconnect.* North Sydney: Bantam.

6 Turkle, S. (2011) *Alone Together: Why We Expect More From Technology and Less From Each Other.* New York: Basic Books.

7 Rosen, L. D. (2010) *Rewired: Understanding the iGeneration and the Way They Learn.* New York: Palgrave Macmillan.

8 Rideout, V. J., Foehr, U. G. and Roberts, D. F., 'Generation M2: Media in the Lives of 8- to 18-Year-Olds', A Kaiser Family Foundation Study, January 2010, http://www.kff.org/entmedia/8010.cfm.

9 Ferriss, T. (2007) *The 4-Hour Workweek*, USA: Crown Publishing Group.

10 Price, I. 'Three cheers for four-day weeks', *The Guardian*, 16 April 2011.

Chapter 10

Opening quote: Mahatma Gandhi.

1 Taylor, J. 'What I learned when I quit Facebook', *The Daily Muse*, 28 February 2012, www.thedailymuse.com/tech/facebook-cleanse/.

2 Hughes, S. 'Cold turkey on a "digital detox"', *BBC News Magazine*, 18 August 2012, http://www.bbc.co.uk/news/magazine-19283726.

3 A list of 'tell all' quitting articles are linked to by Bosker, B. (senior tech editor) 'Why we quit Facebook: The new tell-all', *Huffington Post*, 13 July 2012, http://www.huffingtonpost.com/bianca-bosker/quit-facebook_b_1671872.html.

4 Taylor, J. 'What I learned when I quit Facebook', *The Daily Muse*, 28 February 2012, www.thedailymuse.com/tech/facebook-cleanse/.

5 Hofmann, W., Vohs, K. D. and Baumeister, R. F. 'What people desire, feel conflicted about, and try to resist in everyday life', *Psychological Science*, 2012, 23: 582, http://pss.sagepub.com/content/23/6/582, 11 June 2012 (online first version of record 30 April 2012).

6 Dokoupil, T. 'Is the web driving us mad?', *The Daily Beast*, 9 July 2012.

7 Greenfield, S. (2008) *ID: The Quest for Identity in the 21st Century.* Great Britain: Sceptre.

8 Small, G. 'Techno Addicts', *Psychology Today*, 22 July 2009, www.psychologytoday.com/print/31187.

9 A BT survey in 2012 shows that one third of Britons check work emails on holiday and 40 per cent log in to social networking sites.

10 Abel, J. 'The rise of digital detox', *Forbes*, 15 May 2012, http://www.forbes.com/sites/gyro/2012/05/15/the-rise-of-digital-detox/.

11 Maushart, S. (2010) *The Winter of our Disconnect*. North Sydney: Bantam.

Chapter 11

Opening quote: Honoré, C. (2004) *In Praise of Slow*. London: Orion Books.

1 Evidence that meditation helps restore directed attention abilities comes from Kaplan, S. (2001) 'Meditation, restoration, and the management of mental fatigue', *Environment and Behaviour, 33*, pp. 480–506 and other sources including Tang *et al.*, 2007.

2 Johnstone, M. (2012) *Quiet the Mind*. London: Constable & Robinson Ltd.

3 Alidina, S. (2012) *Relaxation for Dummies*. Chichester: John Wiley & Sons, Ltd.

4 Chatfield, T. (2012) *How to Thrive in the Digital Age*. London: Pan Macmillan.

5 Ibid.

6 Honoré, C. (2004) *In Praise of Slow*. London: Orion Books.

7 Watson, R. (2010) *Future Minds: How the Digital Age is Changing Our Minds, Why This Matters and What We Can Do About It*. London, Boston: Nicholas Brealey Publishing.

8 Berman, M. G, Jonides, J. and Kaplan, S. 'The Cognitive Benefits of Interacting with Nature', *Psychological Science*, Vol. 19, No. 12, December 2008, pp. 1207–12.

9 Kaplan, S. 'The restorative benefits of nature: toward an integrative framework', *Journal of Environmental Psychology*, 15, pp. 169–82, 1995.

10 Berman, M. G, Jonides, J. and Kaplan, S. 'The Cognitive Benefits of Interacting with Nature', *Psychological Science*, Vol. 19, No. 12, December 2008, pp. 1207–12.

Chapter 12

Opening quote: Shlain, T. 'National day of unplugging: a digital detox', *Huffington Post*, 23 March 2012.

1 Maushart, S. (2010) *The Winter of Our Disconnect*. North Sydney: Bantam.

2 Chatfield, T. (2012) *How to Thrive in the Digital Age*. London: Pan Macmillan.

3 Watson, R. (2010) *Future Minds: How the Digital Age is Changing Our Minds, Why This Matters and What We Can Do About It*. London, Boston: Nicholas Brealey Publishing.

4 Joi Ito was speaking in a debate 'Does technology open or close our minds?' on Radio 4, 1 September 2012.

5 Jackson, M. (2008) *Distracted: The Erosion of Attention and the Coming of the Dark Age*. New York: Prometheus Books.

6 Watson, R. (2010) *Future Minds: How the Digital Age is Changing Our Minds, Why This Matters and What We Can Do About It*. London, Boston: Nicholas Brealey Publishing.

7 'Generation IP 2025' by Virgin Media, and a study by Virgin Media and The Future Laboratory: http://www.virginmediabusiness.co.uk/generationip/.

8 'US teen invents advanced cancer test using Google', BBC news website, 21 August 2012, http://www.bbc.co.uk/news/magazine-19291258.

Tuning in to a digital life

Opening quote: Mark Lacas, *SHINE: The Entrepreneur's Journey*, a short film by Dan McComb and Ben Medina, http://biznik.com/SHINE.